D1610361

Neuroscience of Creativity

Neuroscience of Creativity

edited by Oshin Vartanian, Adam S. Bristol, and James C. Kaufman

The MIT Press
Cambridge, Massachusetts
London, England

MIT Press books may be purchased at special quantity discounts for business or sales promotional use. For information, please email special_sales@mitpress.mit.edu.

This book was set in Stone Sans and Stone Serif by Toppan Best-set Premedia Limited, Hong Kong. Printed and bound in the United States of America.

Library of Congress Cataloging-in-Publication Data

Neuroscience of creativity / edited by Oshin Vartanian, Adam S. Bristol, and James C. Kaufman.
pages cm
Includes bibliographical references and index.
ISBN 978-0-262-01958-3 (hardcover : alk. paper)
1. Creative ability. 2. Cognitive neuroscience. I. Vartanian, Oshin, 1970–. II. Bristol, Adam S., 1975–.
BF408.N5196 2013
153.3'5—dc23
2013004389

10 9 8 7 6 5 4 3 2

For Alexandra and Atam

—Oshin Vartanian

For Indre

—Adam S. Bristol

For Avi Ben Zeev,
Chosen brother.

—James C. Kaufman

Contents

Acknowledgments

The authors would like to thank Katie Persons, Judith Feldmann, and Phil Laughlin of the MIT Press for making this process so smooth and simple.

Oshin's work on cognitive training is partly supported by an Applied Research Project funded by Canada's Department of National Defence.

Adam would like to thank Marc Schneidman for his mentorship and for his encouragement of intellectual curiosity among his coworkers. He would also like to thank Indre Viskontas for her unwavering support and guidance.

James would like to thank the people at CSUSB who make his work possible—including (former) President Al Karnig, Provost Andrew Bodman, Dean Jamal Nassar, Chair Robert Ricco, (former) Chair Robert Cramer, Mark Agars, Stacy Brooks, Stephanie Loera, Amanda Ferguson, and the Kaufman Inner Lab Leaders. And, always, Allison.

Introduction

Adam S. Bristol, Oshin Vartanian, and James C. Kaufman

Yet the suns that light the corridors of the universe,
shine dim before the blazing of a single thought.

So proclaims the intrepid Dr. Duval in the 1966 sci-fi film *Fantastic Voyage*, in which a group of scientists miniaturize themselves and enter the brain of a dying patient in search of a cure. What they find is a magical world of unanticipated complexity in an organ that continues to astound its master and servant—animals and man. Of course, modern science's continued effort to understand the neurobiological bases of creativity is its own fantastic voyage, and although we cannot yet insert ourselves into the brains of creative people, technical advances in the neurosciences have begun to reveal the workings of the brain in tremendous detail. With new tools and theories, researchers have generated a body of neuroscientific research and theorizing to complement the nearly one hundred years of sociocultural, cognitive, developmental, educational, and historiometric perspectives on creative processes.

The purpose of this book is to bring together leading researchers from around the world to provide an up-to-date review of empirical and theoretical approaches to the neurobiological bases of creativity. Our hope is that active creativity researchers will find this volume to be a valuable summation of current theoretic and empirical approaches, from which new ideas will be born to inform interdisciplinary perspectives. Our hope is also to inspire advanced students and researchers in adjacent fields for whom creativity is a potential area of focus, and to consider contributing to the advances in creativity research. Indeed, we purposely sought contributions from several scientists who would not call themselves "creativity researchers" per se, but whose research we felt was highly relevant to understanding the neural bases of creativity. The chapters should certainly be engaging for the intelligent and curious lay reader as well.

It would appear that a book on the neuroscience of creativity requires a definition of creativity. However, enforcing a single definition of creativity on the chapters would reflect a rather unnatural representation of the state of affairs in this field, represented as it is by a multitude of context-dependent and domain-specific definitions of what constitutes creativity. Of course, there are some general definitions of creativity that almost everyone in the field agrees on, such as viewing creativity as the *generation of novel and useful products within a specific context* (e.g., Kaufman, 2009; Plucker, Beghetto, & Dow, 2004; Simonton, 2012). We have opted to congregate around this general definition, thereby letting individual researchers define what creativity means in the context of their specific chapters. We believe that this approach more accurately reflects the context- and domain-specific nature of creativity and its neural correlates (Vartanian, 2012).

The book is organized into six sections, which roughly correspond to a progression of theoretical, genetic, structural, clinical, functional, and applied approaches. Admittedly, this is a rather unsatisfactory description of the book's organization as many of the chapters span multiple levels of analysis, which is a testament to the authors and to the demands of addressing the complexities of creativity from a neurobiological perspective. But in general, the chapters progress from the more fundamental levels of genetics and neurophysiology to systems neuroscience and neuroimaging, with important "book ends" that provide theoretical frameworks and synthesis.

The first section of the book, entitled "Theories and Constraints," contains three chapters that provide a context for later discussions. The book begins with a contribution from Margaret A. Boden, who outlines the possibilities and limitations of a neuroscientific approach to the study of creativity. Critically, she notes the distinction between exploratory creativity, transformational creativity, and combinatorial creativity, and highlights the fact that experimental analyses to date have largely concerned themselves with the latter. Her contention is that an understanding of the neurocognitive basis of semantic hierarchies and of knowledge relevance remains elusive, yet is central to creative processes. In chapter 2, Liane Gabora and Apara Ranjan describe a neurally inspired model of creativity that hinges on aspects of memory encoding and retrieval, specifically the reconstructive nature of memory and the novel by-products resulting from contingent activation of subsets of cell assemblies they call "neurds." Finally, W. David Stahlman, Kenneth J. Leising, Dennis Garlick, and Aaron P. Blaisdell, animal behavior experts, draw on fundamental principles of conditioning and associative learning to provide a stimulating overview of what could be viewed as rudimentary creativity in the form of novel and unexplained behavior of animals. Indeed, it appears that there is much to learn about the production of novel behavior from examining nonhuman animals.

The second and third sections of the book, "Genetics" and "Neuropsychology," respectively, contain chapters that explore these areas from multiple perspectives. Chapter 4, by Baptiste Barbot, Mei Tan, and Elena L. Grigorenko, updates the reader on the issue of heritability of creativity and the interaction of genes and environment in shaping creative processes, with a specific focus on the genetic bases underlying the reception of creative products. In chapter 5, Marleen H. M. de Moor, Mark Patrick Roeling, and Dorret I. Boomsma use data from the well-known Netherlands Twin Registry to examine the pattern and degree of talent among parents and their children across multiple artistic and technical domains. Chapter 6, by Indre V. Viskontas and Bruce L. Miller, and chapter 7, by Dahlia W. Zaidel, describe fascinating windows into the brain basis of creativity: neurological findings of altered or, in some cases, emergent artistic talent in brain-damaged patients and patients afflicted with neurodegenerative conditions. Both chapters emphasize the importance of triangulating across multiple approaches for understanding the emergence and alteration of artistic (and aesthetic) abilities.

Next, the book moves to a section entitled "Pharmacology and Psychopathology," discussing areas that have benefited from the continued development of new tools and techniques for interrogating the brain. David Q. Beversdorf reviews work from his lab and others in chapter 8, providing insights into the involvement of multiple neurotransmitter systems in cognition and creative processes. In chapter 9, Shelley Carson provides a review of the association between creativity and mental health, digging deeper into the specific traits and neural mechanisms to account for their shared vulnerabilities and nuanced relationship. Both chapters provide integral information for clinical interventions for creativity.

Neuroimaging, the topic of the book's fifth section, has afforded researchers an extraordinary view into the workings of the human brain in health and disease, and creativity researchers have taken advantage of such tools. In chapter 10, Andreas Fink and Mathias Benedek examine EEG and functional MRI studies of intelligence and creativity, asking to what extent they and their associated neural networks are modifiable by training or influenced by individual differences. In chapter 11, Rex E. Jung and Richard J. Haier provide an integration of many neuroimaging and neuroanatomical studies to propose a new model of neural network functioning across brain regions in which creative processes are dissociable from general intelligence. Both chapters reinforce the importance of continued research into understanding the dissociable and shared neural pathways for intelligence and creativity.

A final section entitled "Aesthetic and Creative Products" focuses on a critical but often overlooked applied aspect of the neurobiological bases of creativity. Chapter 12, by Oshin Vartanian, discusses the use of neuroscientific data as part of the toolkit to enhance creativity within a componential approach. Finally, chapter 13, by Pablo P. L. Tinio and Helmut Leder, steps outside the brain to focus on the nature of art, primarily visual art, and the unique information processing challenges that it poses to the viewer. A thorough examination of this domain, they argue, is a prerequisite to a meaningful exploration of "neuroaesthetics"—a field wherein the generation and appreciation of creative products come full circle.

As we began to receive contributions and the book's contents materialized, we recognized familiar concepts, but also clear illustrations of how the field is progressing. In the end, the book contained not even one reference to the well-trodden Helmholtz model of creative thinking (Helmholtz, 1896, as cited in Wallas, 1926), a depiction of creativity that has inspired hundreds, if not thousands, of spirited discussions and research projects. If the final volume has the impact we anticipate, many new ideas and experiments will be generated from new data and new models of creative processes inspired by it. We hope that creativity researchers, neuroscientists, and scholars in general will find much to ponder in the pages to come.

We would like to thank all of the contributing authors for their thoughtful and timely contributions. We hope you enjoy the book.

ASB, OV, and JCK
September 2012

References

Helmholtz, H. von (1896). *Vortage und Reden*. Brunswick: Friedrich Viewig.

Kaufman, J. C. (2009). *Creativity 101*. New York: Springer.

Plucker, J., Beghetto, R. A., & Dow, G. (2004). Why isn't creativity more important to educational psychologists? Potential, pitfalls, and future directions in creativity research. *Educational Psychologist, 39*, 83–96.

Simonton, D. K. (2012). Taking the U.S. patent office criteria seriously: A quantitative three-criterion creativity definition and its implications. *Creativity Research Journal, 24*, 97–106.

Vartanian, O. (2012). Dissociable neural systems for analogy and metaphor: Implications for the neuroscience of creativity. *British Journal of Psychology, 103*, 302–316.

Wallas, G. (1926). *The art of thought.* New York: Harcourt Brace.

I Theories and Constraints

1 Creativity as a Neuroscientific Mystery

Margaret A. Boden

Just What Sort of Mystery Is This?

Many people, still under the influence of nineteenth-century Romantic views, believe that creativity is a mystery forever beyond the reach of science. They even believe, again echoing Romanticism, that it's a special faculty, confined to a tiny elite. They are wrong. In fact, creativity—the ability to generate ideas/artifacts that are new, surprising, and valuable—is an aspect of human intelligence in general (Boden, 2004, 2010). As such, it's rooted both in our material embodiment and in our sociocultural context—and it depends on the brain. In other words, it's an unsolved puzzle for neuroscientists, not an ineluctable mystery essentially beyond their grasp.

Discussions of creativity are often bedeviled by the discussants adopting different senses of the word "new." An idea may be new to the person who has just come up with it, even though it has occurred to countless other people in the past. Let's call this P-novelty: "P" for psychological. Alternatively, the idea may be new, so far as is known, to the whole of human history. This is H-novelty: "H" for historical. Depending on which sense of "new" is involved, a new idea may or may not count as H-creativity. But it always exemplifies P-creativity, of which H-creativity is clearly a special case. From the psychologist's point of view, and from the neuroscientists' perspective also, the fundamental phenomenon that needs to be explained is P-creativity. Examples of H-creativity—some of which are mentioned below—may be especially interesting to us, as intellectually curious human beings. But they are scientifically relevant primarily as instances of P-creativity: their *historical* situation is not a matter for neuroscience.

If creativity is an unsolved scientific puzzle rather than an occult and ever-enigmatic mystery, it is nevertheless a puzzle that will be very hard to solve. In common parlance, then, it's a "mystery." The air of mystery is strengthened by the fact that introspection rarely helps. Artists, scientists, and mathematicians often report that they have no idea how they came up with their valuable new ideas. Some even use this phenomenological fact to suggest that *they* didn't come up with the novel idea at all: rather, some ultrahuman, perhaps divine, power did so. They forget, of course, that creativity is not unique in this regard: introspection doesn't tell us how we form grammatical sentences, either, nor how we interpret photographs as depicting specific scenes. In general, much more goes on in our minds below the level of consciousness than can ever be accessed by it. Were that not so, we'd be paralyzed by information overload. Psychology faces "introspective mystery" in all areas of mental life.

The main difficulty in solving the puzzle of creativity is not—as is also widely believed—that it is unpredictable. Creativity is indeed largely unpredictable, for a number of different reasons (Boden, 2004, chapter 9). The most important reason is the enormous complexity, and idiosyncrasy, of human minds, the detailed contents of which are largely unknown even to the individual concerned. Marcel Proust himself couldn't have predicted that a flood of memories would be prompted by his eating the famous madeleine. As for third-party observers, even if (which is unlikely) someone had happened to know that he used to eat madeleines as a youth at his grandmother's house, they too would have been unable to predict the host of mental associations that were triggered by his eating them again in adult life. Even if only one thought is of interest, psychological complexity may hide it from view: the very best clinical psychologist may not know whether or not Joe Bloggs will decide to commit suicide—still less just when and how.

In one sense, this does put creativity outside the scope of science. However, that's no reason for the scientist to despair—and no reason to mark creativity off from other, notionally less mysterious, phenomena. For it's not the aim of science to predict individual events—most of which, unlike Joe Bloggs's suicide, are of no interest to us, anyway: we don't *want* physicists to be able to predict the movements of each grain of sand on the beach. Even if the suicidal thoughts are assigned some statistical *probability*, this may not be calculated on purely scientific grounds (see Meehl, 1954.) Occasionally, events can be precisely predicted by science: think of a returning space-capsule, splashing into the Pacific Ocean with rescue-ships already waiting nearby. Usually, however, they cannot. Science in general isn't focused on the prediction of particularities, even though prediction is an important aspect of the experimental method. Rather, it seeks to show how events of a certain class are *possible*, and how they are related to other sorts of events, whether actual or merely conceivable (Boden, 2006: 7.iii.d).

Accordingly, a neuroscientific explanation of the puzzling phenomenon of creativity would show us *how it is possible* for this still-mysterious phenomenon to occur. The common view that a science of creativity could predict every detail of creative thought, thus making human artists and scientists (and everyday punsters ...) redundant, is mistaken. The "mystery" of creativity, as regards neuroscience, lies not in its unpredictability but in its computational variety. As outlined below, there are several different types of creativity, involving distinct sorts of information processing. A satisfactory neuroscience of creativity would have to illuminate each one of these. "Illumination," here, means significantly more than locating the brain areas involved. In general, a neuroscientific *explanation* of a psychological phenomenon does not merely tell us which parts of the brain, and/or which neuronal groups, are active when the phenomenon occurs. Crucially, it tells us *what the brain cells are doing,* where this is understood not in terms of (for instance) chemical changes but in terms of the computations, or information processing, that the cells are performing (Boden, 2006, chapter 14).

The computational psychologist John Mayhew, when explaining stereopsis, put it like this: "Finding a cell that recognizes one's grandmother does not tell you very much more than you started with; after all, you know you can recognize your grandmother. What is needed is an answer to how you, or a cell, or anything at all, does it. The discovery of the cell tells one what does it, but not how it can be done" (Mayhew, 1983, p. 214). Even if the detailed neuronal circuits involved are known, *what the circuits are doing* may be obscure. The key questions concern what information is received and/or passed on by the cell or cell group, and how it's computed by them. Put another way, they concern "how electrical and chemical signals are used in the brain to represent and process information" (Koch & Segev, 1989, p. 1). The key point of this chapter, then, is that we need to know *what sort of information processing* is involved in creativity, to have any hope of a neuroscientific explanation of it. And the conclusion will be that we are at present within reach of such an explanation only for one type of creativity. The others will be much more difficult nuts for the neuroscientist to crack.

The Three Types of Creativity

Creativity can happen in three main ways, only one of which is typically recognized by people trying to analyze it (including those experimental psychologists who specialize in this area). Specifically, creativity may be combinational, exploratory, or transformational (Boden, 2004, chapters 3–6). These are distinguished by the sorts of psychological process that are involved in generating the new idea. A satisfactory neuroscientific theory of creativity would need to explain how each of the three types can come about.

Combinational creativity—which is usually the only type recognized in studies/definitions of creativity—involves the generation of unfamiliar combinations of familiar ideas. In general, it gives rise to a "statistical" form of surprise, like that experienced when an outsider wins the Derby. Everyday examples of combinational creativity include visual collage (in advertisements and MTV videos, for instance); much poetic imagery; all types of analogy (verbal, visual, or musical); and the unexpected juxtapositions of ideas found in political cartoons in newspapers. Scientific examples include seeing the heart as a pump, or the atom as a solar system.

Exploratory and transformational creativity are different. Unlike the combinational variety, they're both grounded in some previously existing, and culturally accepted, structured style of thinking, or "conceptual space." Of course, combinational creativity, too, depends on a shared conceptual base—but this is, potentially, the entire range of concepts and world-knowledge in someone's mind. A conceptual space, or thinking style, is both more limited and more tightly structured (often, hierarchically). It may be a board game, for example (chess or Go, perhaps), or a class of chemical structures (aromatic molecules, for instance), or a particular type of music or sculpture.

In exploratory creativity, the existing stylistic rules or conventions are used to generate novel structures (ideas or artifacts), whose possibility may or may not have been realized before the exploration took place. To the extent that it was not, the new structure will be not only satisfying but surprising. A new painting in the impressionist style, a new benzene derivative, or a new fugue or sonnet are all examples. So is the daily generation of new sentences, fitting the grammatical rules of the language in question. Exploratory creativity can also involve the search for, and testing of, the specific stylistic limits concerned. Just which types of structure can be generated within this space, and which cannot?

Transformational creativity is the most arresting of the three. Indeed, it leads to "impossibilist" surprise, wherein the novel idea appears to be not merely new, not even merely strange, but *impossible*. Seemingly, it simply could not have arisen—and yet it did. In such cases, the shocking new idea arose because some defining dimension of the style, or conceptual space, was altered—so that structures can now be generated that *could not* be generated before. The greater the alteration, and the more fundamental the stylistic dimension concerned, the greater the shock of impossibilist surprise.

For instance, imagine altering the rule of chess that says that pawns cannot jump over other pieces: they're now allowed to do this, as knights always were. The result would be that some games of chess could now be played that were literally *impossible* before. Or consider the suggestion, new in 1865, that the benzene molecule may be a ring of carbon atoms: a topologically closed string, rather than—like all previously described molecules—an open one. *Exploratory* creativity then took over, as organic chemists mapped the space of benzene derivatives. They later went on to ask whether the core of some ring-molecules might include five atoms rather than six, and/or atoms of elements other than carbon. Whether one chooses to call those two questions "exploratory" or "transformational" is negotiable. The important point is that they were both driven by specific features of the benzene space that had been explored for some time.

A comparable, and much more recent, example concerns the shocking idea that some carbon molecules may be hollow spheres. The key transformation, here, was to consider atomic bonds forming not just in one spatial dimension (as in a planar sheet of graphene), but in three. What's generally regarded as the key paper was published in 1985 (Kroto et al., 1985). It reported experimental research on carbon vapors heated to thousands of degrees, in which various multiatom molecules (but mostly the soccer-ball C60, or buckminsterfullerene) formed spontaneously. Subsequent *exploratory* creativity synthesized many new "fullerenes" of differing shapes and sizes. These included open-ended or closed tubes (formed when a few percent of nickel or cobalt atoms were added) that could act as molecule-carriers and electronic conductors, so providing for a host of novel technological applications. This pioneering work led to a Nobel Prize eleven years later (Smalley, 1996).

That work was rightly seen by the Nobel committee as "pioneering," not least because of its detail and systematicity (made possible by the team's development of laser-instrumentation for measurement). In fact, however, the central "shocking idea" had been suggested in 1970, by chemists in Japan and in the UK. But it was then considered too bizarre to be accepted (valued) by the scientific community. Moreover, a closely similar idea, envisaging the addition of impurities to a planar network of carbon atoms (and soon pointing out that the resultant hollow molecules might carry other molecules inside them), had been published in the *New Scientist* as early as 1966—but the author had presented this as scientific fantasy rather than serious research (Jones, 1966; cf. 1982, pp. 118–119).

This example illustrates the difficulty, in many cases, of deciding whether a particular idea really is new, and/or really is valuable (see below).

In general (though less so in literature), transformational creativity is esteemed more highly than the other two varieties. The people whose names are recorded in the history books are usually remembered above all for changing the accepted style. Typically, the stylistic change meets initial resistance. And it often takes some time to be accepted. That's no wonder; transformational creativity *by definition* involves the breaking/ignoring of culturally sanctioned rules.

However, novel transformations are relatively rare. All artists and scientists spend most of their working time engaged in combinational and/or exploratory creativity. That's abundantly clear when one visits a painter's retrospective exhibition, especially if the canvasses are displayed chronologically: one sees a certain style being adopted, and then explored, clarified, and tested. It may be superficially tweaked (a different palette adopted, for example). But it's only rarely that one sees a radical transformation taking place. Similarly, the list of a scientist's research papers rarely includes a transformative contribution: mostly, scientists explore the implications of some already accepted idea. Even if that idea is itself transformative, and relatively recent, it normally prompts exploration rather than further transformation. That was so in the case of ring-molecules, as we've seen; and the case-history of the fullerenes provides further illustrations. Only very seldom does an individual scientist, or artist, make more than one transformative move. Picasso is an example from the arts of someone who pioneered several distinct styles over his lifetime. In science, the Crick–Watson team discovered both the double helix and, a few years later, the genetic code.

The saga of the fullerenes also illustrates the fact that identifying a "creative" idea, or a scientific "discovery," is not always straightforward. Such judgments can even be affected by national rivalries, not to mention social snobbery and personal jealousies (Schaffer, 1994). The identification of creativity is *never* purely scientific. Even though science can occasionally explain why we have certain values (shininess, for instance—see Boden, 2006: 8.iv.c), it cannot, in principle, *justify* any value. Moreover, our values often change: different social groups/subgroups, in differing times and places, may value very different things. Because the notion of *positive valuation* is included within the concept of creativity, the class of "creative" ideas is not a natural kind. In other words, it is not a purely scientific concept. It follows that neuroscience could never explain the origin of creative ideas without some prior (socially based) judgments identifying *these* ideas as creative, in contrast with others that are merely new. Even novelty isn't always easily judged, as the case history of the fullerenes shows (see Boden, 2006: 1.iii.f–g).

A final complication must be mentioned here. Namely, what we naturally think of as a "single" idea or artifact may involve more than one sort of creativity. The three forms of creativity distinguished above are analytically distinct, in that they involve different types of psychological processes for generating novel ideas. But a given artwork or scientific theory can involve more than one type. That's partly why it's generally more sensible to ask whether this or that *aspect* of the idea in question is creative, and in what way. A neuroscientific theory of creativity should be able to show how the three forms of creativity can be integrated, as well as how they can function independently.

What Might Neuroscience Have to Say?

There's no doubt that neuroscience could help to show how combinational creativity is possible. Indeed, it already has. Neurological studies and computer models of associative memory have already thrown light on the mechanisms underlying much poetic imagery. The richness and subtlety of these associations have long been appreciated by literary scholars. The best example, here, is John Livingstone Lowes's (1930) masterly literary detective story tracing the detailed origins of Samuel Taylor Coleridge's imagery in *The Rime of the Ancient Mariner* and *Kubla Khan* (Boden, 2004, chapter 6). In relation to the pessimism about particularism expressed earlier, it's worth mentioning that Livingston Lowes had access not only to the whole of Coleridge's eclectic library but also to his commonplace books for the eighteen months during which these poems were written, in which he had jotted down quotations that had interested him. That degree of access to the detailed contents of another person's mind is highly unusual.

However, beyond the already long-familiar idea that brains are composed of interconnected units that are somehow responsible for conceptual associations (Hartley, 1749), Lowes knew nothing of the neural mechanisms involved. Today, we are in a very different position. It was known by the 1980s that certain drugs can increase or decrease the associative range of conceptual thinking, leading to more or less inclusive and/or idiosyncratic combinations, respectively (e.g., Shaw et al., 1986; cf. Eysenck 1994, pp. 224–232). And now, we have much more data, and many more neuroscientific (not least neurocomputational) concepts, to work with.

This isn't to say that we can now come closer to literary particularism than Livingston Lowes, for instance, could. In other words, it's not to say that neuroscience could ever explain just how/why *this* idea was associated with *that* idea on a given occasion. Even if the idea in question could be neuronally located, as intentional verbs, for instance, have been located in the posterior superior temporal sulcus (pSTS, Allison et al., 2000; Castelli et al., 2002; Frith & Frith, 2003), the specific association that arose in some individual's mind could not be explained in detail—still less predicted. However, we saw earlier that particularist explanation/prediction is not the aim of science. Insofar as such particularist insights are available they are post hoc, not predictive, and are to be found rather in the humanities (Livingstone Lowes's discussion of *The Rime of the Ancient Mariner* provides some exceptionally convincing examples).

Associative pathways, however, are not all there is to combinational creativity. There is also the tricky issue of *relevance*. Conceivably, any concept could be associated with any other, by some sufficiently tortuous neuronal path. In that sense, there's no limit to the number of "unfamiliar combinations" that are possible. But life is too short to follow only highly tortuous pathways. Even poets have to provide enough context to make their meaning communicable; and everyday speech, in general, has to be understood *immediately*. In other words, those novel combinations that we *value*, and thus regard as "creative," invariably involve relevance—even if the relevance is not immediately apparent.

An insightful computational approach to relevance suggests that we have evolved an involuntary, and exceptionless, principle of communication (and problem solving) based on a cost–benefit analysis, weighing effort against effect (Sperber & Wilson, 1986). The more information-processing effort it would take to bear x in mind in the context of y, the more costly this would be: and high cost gives low relevance. The more implications (regarding things of interest to the individual concerned) that would follow from considering x, the more effective it would be: and high effectiveness gives high relevance.

The suggestion here is not (paradoxically) that we precompute just what effort/effect would be involved in considering a certain concept. Rather, there must be psychological mechanisms evolved for recognizing relevance. For example, our attention is naturally caught by movement, because moving things are often of interest. Similarly, even a newborn baby's attention is preferentially caught by human speech sounds. Besides being built into our sensory systems, relevance recognition is built into our memories: it's no accident, on this view, that similar and/or frequently co-occurring memories are easily accessible, as they are "stored" together in scripts, schemas, and conceptual hierarchies.

Different cognitive strategies may vary in the measure of cost or benefit that they attach to a given conceptual "distance." Surrealists, for example, tolerate greater distances than straightforwardly "representational" writers and painters do—hence the extreme unfamiliarity of the novel combinations found in their work. The artist's personal signature, which can affect many aspects of a creative work (see Boden, 2010), can apply here: one individual surrealist may be even more forgiving of conceptual distance than another. Similarly, different rhetorical styles in literature involve different levels of cost and/or different types of information processing in both writer and reader: compare Charles Dickens and James Joyce, for instance. A literary personal signature may also involve a preference for finding many sorts of relevance in certain concepts: *animals*, for the poet Ted Hughes, for example.

This analysis of relevance implies that, *pace* symbolic computationalists such as Jerry Fodor (1983), labored scientific inference is *not* a good model for everyday, instantaneous understanding (Sperber & Wilson, 1986). Similarly, it rejects the GOFAI (Good Old-Fashioned Artificial Intelligence) assumption that deliberate reasoning (which is needed by literary scholars and historians when puzzling over obscure texts) is required for spontaneous interpretation (Sperber & Wilson, 1986, p. 75). Rather, our understanding typically depends on associative, nonlogical guessing that is constrained by what we take to be relevant.

It follows that a satisfactory neuroscientific account of combinational creativity would identify the various mechanisms evolved for judging relevance. Given that this matter is a verbal/conceptual version of the notorious frame problem (see McCarthy & Hayes, 1969; Sperber & Wilson, 1996; Boden, 2006, pp. 771–775, 1003–1005), that is a tall order. With respect to the other two forms of creativity, there's more bad news: they are significantly less amenable to neuroscience. That's true in two ways. First, we rarely know all the constraints defining the conceptual spaces of art or science, still less the computational processes required to explore and/or to transform them. Historians of art and musicologists spend lifetimes in attempting to make stylistic constraints explicit, and succeed only to a very limited degree. Sometimes, they even announce a given style to be unfathomable. For instance, an architectural historian specializing in Frank Lloyd Wright's work announced the style (the principle of "balance") of his Prairie Houses to be "occult" (Hitchcock, 1942).

One of the advantages of computer modeling is that it can sometimes help to develop, and to test, explicit theories about such matters. So, for instance, a computerized "shape grammar" has generated every one of

Wright's forty-or-so Prairie House designs, plus many others clearly sharing the same style—without ever producing one that lacks this intuitively recognizable principle of unity (Koning & Eizenberg, 1981). Moreover, this work has shown that the *fireplace* is key to the style. That is, when generating specific design choices, changes to the location of the fireplace (or to the number of fireplaces) result in changes to most other aspects of the house.

The second type of "bad news" is that, even if we had defined the conceptual spaces concerned, and even if we knew the generative processes involved in negotiating and changing them, we wouldn't know how these are neurally embodied. We might assign them to some central cognitive workspace (e.g., Baars, 1988; Changeux, 2002), to be sure. And we might even be able to locate that workspace, very broadly, in the brain. But knowing just how sonnet-form, for instance, is neurally embodied, and how it is neurophysiologically accessed in generating "Shall I compare thee to a summer's day?," is way beyond the state of the art.

This is not just a difficulty in particularistic prediction, as discussed above: rather, it's a difficulty in knowing *how it is possible* for neurological mechanisms to implement sonnet-form, and to exploit it so as to generate the line in question. Similarly, explaining—in neurological terms—just how the Prairie House style can generate the Henderson house, the Martin house, or the Baker house (different examples, each named after the clients who commissioned them) is at present beyond us.

My own view is that such explanations are likely to remain beyond us for very many years, perhaps even forever. That's not because I agree with those philosophers (e.g., McGinn, 1989, 1991) who argue that the explanation of high-level thought and consciousness is as far beyond the cognitive capacities of *Homo sapiens* as theoretical physics is beyond the capacities of squirrels and chimpanzees. I believe that position to be unnecessarily defeatist. Nevertheless, there are some fundamental problems here, which can't be solved by (theory-free) correlative brain-imaging, or by reference, for example, to trial-and-error combinations and neural evolution (Changeux, 1994).

One of these problems concerns the neural implementation of hierarchy. Most of the styles, or conceptual spaces, explored in art and science are hierarchical. The Prairie House fireplace, for instance, is key to the genre because it lies at a fundamental level in the stylistic hierarchy (the "space grammar") concerned, so that a decision about the fireplace will constrain many later decisions about other, superficially unrelated matters. And the generation of "Shall I compare thee to a summer's day?" requires exploration of grammatical hierarchy. At present, we have no good ideas about how conceptual hierarchies are neurally embodied, or how they can be rationally negotiated in creative thinking.

Still less do we know how transformational procedures may be embodied that can alter those hierarchies. Even domain-general transformations (such as *consider the negative* or *drop a constraint*) are a mystery. And the neural basis of the many domain-specific procedures that led from early Renaissance music (broadly: one composition, one key), through increasingly daring modulations and harmonies, to atonal music is even more elusive (Boden, 2004, pp. 71–74; Rosen, 1976).

One might suggest, at this point, that computer simulation could help. And in principle, it could. However, a *neuroscientifically* plausible model is going to be connectionist rather than symbolic. Yet only symbolic models (a.k.a. GOFAI, Haugeland, 1985, p. 112) are well suited to represent hierarchy. Connectionist models, in general, are not. Despite heroic efforts in that direction, this problem has not yet been solved (Boden, 2006: 12.viii). Perhaps the most impressive attempt is harmony theory (Smolensky et al., 1993; Smolensky & Legendre, 2006), which draws on neuroscientific knowledge. However, this was developed specifically to deal with grammatical hierarchy (syntax), and it's not clear how it could be generalized to model conceptual hierarchies such as artistic or scientific styles. Even if it could, there would be a huge gap between harmony-theoretic modeling and the neurological reality. Most connectionist models, *especially* those intended as models of psychological (not just neurological) functions, rely on computational units that—as compared with real neurons—are too neat, too simple, too few, and too "dry" (Boden, 2006: 14.ii). In brief, the networks studied by connectionist AI are highly nonneural nets.

To be sure, connectionism is becoming gradually more realistic. One recent textbook, featuring the *Leabra* software system developed by its authors, makes great efforts to integrate connectionist AI with neuroscience (O'Reilly & Munakata, 2000). For example, the activation function controlling the spiking of the simulated neurons in *Leabra* is only "occasionally" drawn from mathematical connectionism (ibid., p. 42). Usually, it is based on facts about the biological machinery for producing a spike, including detailed data on ion channels, membrane potentials, conductance, leakages, and other electrical properties of nerve cells (pp. 32–48). Similarly, the basic equations used by *Leabra* when simulating high-level phenomena such as reading or conceptual memory are (usually) painstakingly drawn from detailed biophysical data. This is true, for example, of the equation used for integrating many inputs into a single neuron (see O'Reilly & Munakata's explanation of equation 2.8 on p. 37).

The *Leabra* authors drew the line at applying this equation "at every point along the dendrites and cell body of the neuron, along with additional equations that specify how the membrane potential spreads along neighboring points of the neuron" (p. 38). They had no wish "to implement hundreds or thousands of equations to implement a single neuron," so used an approximating equation instead. But, characteristically, they provided references to other books that did explain how to implement such detailed single-neuron simulations. In general, the psychological models developed by O'Reilly and Munakata would have been different had the neuroscientific data been different. Their discussion of dyslexia, for instance, built not only on previous connectionist work (e.g., Plaut & Shallice, 1993; Plaut et al., 1996), but also on recent clinical and neurological information (O'Reilly & Munakata, 2000, pp. 331–341). As our knowledge of the brain advances, future psychological models—they believe—will, or anyway should, be different again. They see their book as "a 'first draft' of a coherent framework for computational cognitive neuroscience" (ibid., p. 11). With respect to creativity, this implies that we may hope for future connectionist models that embody specific neuroscientific data *as well as* a better understanding of the complex computational processes involved in all three types of creativity. But to hope is not to have (and I'm not holding my breath).

Wittgenstein and Neuroscience

I've assumed so far that it is *coherent* to aim for a neuroscientific explanation of creativity—and, for that matter, of any other psychological phenomenon. In other words, such an explanation is possible in principle, irrespective of whether it has been, or is ever likely to be, achieved. And I've written as though the only reason for denying this is the mysterian view that there is something essentially quasi-magical about creativity, which puts it beyond the reach of science.

However, many philosophers of mind would deny the possibility of a scientific understanding of creativity—and of any other psychological phenomenon—on very different grounds. These writers include the followers of Ludwig Wittgenstein, who suggested in his *Philosophical Investigations* (1953) that there is no level of psychological explanation between remarks about conscious phenomenology and observations about the physical mechanisms of the brain. So, for instance, Richard Rorty explicitly looked forward to "the disappearance of psychology as a discipline distinct from neurology" (1979, p. 121).

Wittgensteinians in general reject psychological explanations posed at the subpersonal level, and thus criticize those *neuroscientific* theories that define brain-processes cognitively (or computationally), rather than purely neurologically. They accuse neuroscientists of incoherence due to the "mereological fallacy," which is to attribute to a part of a system some predicate that is properly attributed only to the whole (Bennett & Hacker, 2003). In this context, the "system" in question is the whole person, the "parts" are the brain (or parts thereof), and the "predicates" are psychological terms such as knowledge, memory, belief, reasoning, choice—and, of course, creativity.

On this view, there is absolutely no hope of a naturalistic psychology. Insofar as psychology exists as a scientific discipline it is said to be a hermeneutic, not a natural science (cf. Harre, 2002; McDowell, 1994). So neuroscience could never *replace* psychology, in the sense of substituting for it. At most, a (noncognitivist) neuroscience could compensate for the lack of a cognitivist (subpersonal) psychology. This rejection of naturalism in psychology reflects a deep divide in Western philosophy, which we can't go into here (but see Boden, 2006: 16.vi–viii). A few neuroscientists (such as followers of Humberto Maturana and Francisco Varela, 1980) lie on the antinaturalist side of the divide. But the vast majority do not. Moreover, neuroscience itself has become increasingly cognitivist—indeed, computational—since the 1950s (Boden, 2006, chapter 14). Information processes and computational mechanisms are now considered crucial in many neuroscientific explanations, from studies of vision to the higher thought processes. And this essay has argued that the computational level of theorizing is crucial in explaining creativity, too. So although Wittgenstein might seem at first sight to be the neuroscientist's friend, perhaps he is not such a good friend after all. Not a false friend, to be sure (for that would involve insincerity or betrayal). But, in my view, a mistaken one.

Conclusion

Nothing that's been said above suggests that there can never be a neuroscience of creativity. Indeed, a neuroscience of *combinational* creativity is arguably within sight—if not yet within reach. It's not yet in reach, partly because—as explained above—challenging problems remain concerning how we make judgments of *relevance* when engaging in, or appreciating, combinational creativity. A neuroscientific explanation of that is not within sight. Moreover, given that this is a verbal/conceptual version of the notorious frame problem (Boden, 2006, pp. 771–775, 1003–1005; Sperber & Wilson, 1996), a neuroscientific explanation is a tall order.

Further reasons why a neuroscience of creativity is not within reach involve hierarchy, as we've seen. Clearly, it must be possible, somehow, for hierarchy—and all other aspects of symbolic thinking—to be implemented in (broadly) connectionist systems (see Gabora & Ranjan, this vol.). After all, the human brain is such a system. However, we need to understand, much better than we do at present, how a basically connectionist system can emulate a symbolic one (i.e., how connectionism can emulate a von Neumann machine).

In addition to highly general questions such as that one, we need to focus on the specific structure of, and the generative processes within, the myriad conceptual spaces underlying science and art. For neither exploratory nor transformational creativity can be properly understood without taking those computational features into account.

References

Allison, T., Puce, A., & McCarthy, G. (2000). Social perception from visual cues: Role of the STS region. *Trends in Cognitive Sciences, 4,* 267–278.

Baars, B. J. (1988). *A cognitive theory of consciousness.* Cambridge: Cambridge University Press.

Bennett, M. R., & Hacker, P. M. S. (2003). *Philosophical foundations of neuroscience.* Oxford: Blackwell.

Boden, M. A. (2004). *The creative mind: Myths and mechanisms* (2nd Ed.). London: Routledge.

Boden, M. A. (2006). *Mind as machine: A history of cognitive science.* Oxford: Clarendon Press.

Boden, M. A. (2010). Personal signatures in art. In M. A. Boden, *Creativity and art: Three roads to surprise* (pp. 92–124). Oxford: Oxford University Press.

Castelli, F., Frith, C. D., Happe, F., & Frith, U. (2002). Autism, Asperger Syndrome, and brain mechanisms for the attribution of mental states to animated shapes. *Brain, 125,* 1839–1849.

Changeux, J.-P. (1994). Creative processes: Art and neuroscience. *Leonardo, 27,* 189–201.

Changeux, J.-P. (2002). *The physiology of truth: Neuroscience and human knowledge* (DeBevoise, M. B., Trans.). Cambridge, MA: Harvard University Press.

Eysenck, H. J. (1994). The measurement of creativity. In M. A. Boden (Ed.), *Dimensions of creativity* (pp. 199–242). Cambridge, MA: MIT Press.

Fodor, J. A. (1983). *The modularity of mind: An essay in faculty psychology*. Cambridge, MA: MIT Press.

Frith, U., & Frith, C. D. (2003). Development and neurophysiology of mentalizing. *Philosophical Transactions of the Royal Society of London, Series B: Biological Sciences, 358*, 459–473.

Harre, R. M. (2002). *Cognitive science: A philosophical introduction*. London: Sage.

Hartley, D. (1749). *Observations on man: His frame, his duty, and his expectations. London*. (Facsimile reproduction ed. T. L. Huguelet, reprinted Gainesville, Florida: Scholars' Facsimiles and Reprints, 1966.)

Haugeland, J. (1985). *Artificial intelligence: The very idea*. Cambridge, MA: MIT Press.

Hitchcock, H. R. (1942). *In the nature of materials: The buildings of Frank Lloyd Wright, 1887–1941*. New York: Meredith Press.

Jones, D. E. H. [as *Daedalus*]. (1966). Note in Ariadne column. *New Scientist, 32*, 245.

Jones, D. E. H. (1982). *The inventions of Daedalus*. Oxford: W. H. Freeman.

Koch, C., & Segev, I. (Eds.). (1989). *Methods in neuronal modeling: From synapses to networks*. Cambridge, MA: MIT Press.

Koning, H., & Eizenberg, J. (1981). The language of the prairie: Frank Lloyd Wright's prairie houses. *Environment and Planning, B, 8*, 295–323.

Kroto, H. W., Heath, J. R., O'Brien, S. C., Curl, R. F., & Smalley, R. E. (1985). C60: Buckminster fullerene. *Nature, 318*, 162–163.

Livingston Lowes, J. (1930). *The road to Xanadu: A study in the ways of the imagination* (Rev. Ed., 1951). London: Houghton.

McCarthy, J., & Hayes, P. J. (1969). Some philosophical problems from the standpoint of artificial intelligence. *Machine Intelligence, 4*, 463–502.

McDowell, J. (1994). *Mind and world*. Cambridge, MA: Harvard University Press.

McGinn, C. (1989). Can we solve the mind-body problem? *Mind, 98*, 349–366.

McGinn, C. (1991). *The problem of consciousness*. Oxford: Blackwell.

Maturana, H. R., & Varela, F. J. (1980). *Autopoiesis and cognition: The realization of the living*. Boston: Reidel. (First published in Spanish, 1972.)

Mayhew, J. E. W. (1983). Stereopsis. In O. J. Bradick & A. C. Sleigh (Eds.), *Physical and biological processing of images* (pp. 204–216). New York: Springer-Verlag.

Meehl, P. E. (1954). *Clinical versus statistical prediction: A theoretical analysis and a review of the evidence*. Minneapolis: University of Minnesota Press.

O'Reilly, R. C., & Munakata, Y. (2000). *Computational explorations in cognitive neuroscience: Understanding the mind by simulating the brain.* Cambridge, MA: MIT Press.

Plaut, D. C., McClelland, J. L., Seidenberg, M. S., & Patterson, K. (1996). Understanding normal and impaired word reading: Computational principles in quasi-regular domains. *Psychological Review, 103,* 56–115.

Plaut, D., & Shallice, T. (1993). Deep dyslexia: A case study of connectionist neuropsychology. *Cognitive Neuropsychology, 10,* 377–500.

Rorty, R. (1979). *Philosophy and the mirror of nature.* Princeton, NJ: Princeton University Press.

Rosen, C. (1976). *Schoenberg.* Glasgow: Collins.

Schaffer, S. (1994). Making up discovery. In M. A. Boden (Ed.), *Dimensions of creativity* (pp. 14–51). Cambridge, MA: MIT Press.

Shaw, E. D., Mann, J. J., & Stokes, P. E. (1986). Effects of lithium carbonate on creativity in bipolar outpatients. *American Journal of Psychiatry, 143,* 1166–1169.

Smalley, R. E. (1996). Discovering the fullerenes. [Nobel Prize acceptance speech.] *Reviews of Modern Physics, 69,* 723–730.

Smolensky, P., & Legendre, G. (2006). *The harmonic mind: From neural computation to optimality-theoretic grammar.* Cambridge, MA: MIT Press.

Smolensky, P., Legnendre, G., & Miyata, Y. (1993). Integrating connectionist and symbolic computation for the theory of language. *Current Science, 64,* 381–391.

Sperber, D., & Wilson, D. (1986). *Relevance: Communication and cognition.* Oxford: Blackwell.

Sperber, D., & Wilson, D. (1996). Fodor's frame problem and relevance theory. *Behavioral and Brain Sciences, 19,* 530–532.

Wittgenstein, L. (1953). *Philosophical investigations* (Anscombe, G. E. M., Trans.). Oxford: Blackwell.

2 How Insight Emerges in a Distributed, Content-Addressable Memory

Liane Gabora and Apara Ranjan

We begin this chapter with the bold claim that it provides a neuroscientific explanation of the magic of creativity. Creativity presents a formidable challenge for neuroscience. Neuroscientific research generally involves studying what happens in the brain when someone engages in a task that involves responding to a stimulus, or retrieving information from memory and using it the right way or at the right time. If the relevant information is not already encoded in memory, the task generally requires that the individual make systematic use of information that *is* encoded in memory. But creativity is different. It paradoxically involves studying how someone pulls out of his or her brain something that was never put into it! Moreover, it must be something both new and useful, or appropriate to the task at hand. The ability to pull out of memory something new and appropriate that was never stored there in the first place is what we refer to as the *magic of creativity*. We will see that (like all magic acts) it isn't really magic after all; there is a clever trick behind it.

The difficulty of achieving a neuroscientific account of creativity goes far beyond the problem of getting people to be creative on demand. Even if we are so fortunate as to determine which areas of the brain are active and how these areas interact during creative thought, we will not have an answer to the question of how the brain comes up with creative solutions and products that are new and appropriate. Although standard technologies for investigating brain activity such as fMRI, EEG, and PET may have much to tell us about creativity, in their current state of development they focus at too high a level to explain the magic of creativity.

On the other hand, since the representational capacity of neurons emerges at a level that is higher than that of the individual neurons themselves—there is no "grandmother neuron" that always responds to your grandmother, or to Halle Berry, and nothing else—the inner workings of neurons take place at too low a level to explain the magic of creativity.

Although it is a little unsatisfying, even if we do not yet know everything there is to know about the conditions under which individual neurons form new dendrites, or what temporary or permanent intracellular changes take place in response to novel stimuli, to make headway we must assume these things happen *somehow*, and move up a level.

Hence we look to a level that is midway between gross brain regions and neurons. Since creativity generally involves combining concepts from different domains, or seeing old ideas from new perspectives, we focus our efforts on the neural mechanisms underlying the representation of concepts and ideas. Thus, we ask questions about the brain at the level that accounts for its representational capacity: at the level of distributed aggregates of neurons (Blasdel & Salama, 1986; Chandrasekharan, 2009; Churchland & Sejnowski, 1992; Dayan & Abbott, 2001; Eliasmith & Anderson, 2003; Hebb, 1949, 1980; Lin et al., 2005; Lin, Osan, & Tsien, 2006; Smith & Kosslyn, 2007).

We lack methods that would permit the necessary spatial resolution to zero in on distributed aggregates of neurons as someone creates, and the necessary temporal resolution to see what these neurons are doing. So in choosing to focus on this level we are, in a sense, groping in the dark. We focus on this level not because the proper tools already exist—not because that's where the light is—but because that's where we need to look—that's where we think the quarter is. In the absence of the ability to directly manipulate and experiment with events at the level midway between gross brain regions and neurons during real *in vivo* bouts of creative thinking, computational models, and a little detective work, play roles in our account of the mechanisms underlying creative insight. We believe that we have located something that has the size, shape, and feel of the real quarter. It explains what is commonly referred to as "thinking outside the box," or what Boden (1990) refers to as *transformational creativity*: creativity that involves not just exploring but changing the space of possibilities. The bulk of this account was put forward over a decade ago (Gabora, 2000) and refined since (Gabora, 2001, 2002, 2010); here we summarize the key elements and frame them in terms of more recent neuroscientific findings.

The Ingenious Way That Representations Are Encoded in Memory

We said that creativity involves pulling something out of your brain that was never put into it. Nevertheless, it is generally assumed that what gets pulled out bears *some* relationship to knowledge and experiences encoded in memory before the creative act took place. But although we spend 15 to 50 percent of our time engaged in "mindwandering," recalling and playing with existing knowledge and ideas in an undirected manner (Smallwood & Schooler, 2006), little of this is what we would call creative; only rarely would one of the thoughts one has during a bout of mindwandering qualify as an insight (Andrews-Hanna, Reilder, Huang, & Buckner, 2010; Signer & Antrobus, 1963). So one question is: what is going on above and beyond the usual tinkering and rearranging that occurs in everyday mindwandering when a genuinely creative idea emerges?

A second question is the following. There are potentially infinitely many different ways of tweaking what we know to come up with something new. How is it that people so often manage to hit on ideas that are *just right*? We believe that the answer to these questions can be obtained by looking at the ingenious way that one's history of experiences is encoded in memory. A brain contains information that was never explicitly stored there but that is *implicitly* present nonetheless. We propose that this implicitly present information enables one to go beyond what one knows without resorting to trial and error.

We take as a starting point some fairly well-established characteristics of memory. Human memories are encoded in neurons that are sensitive to ranges (or values) of *microfeatures* (Churchland & Sejnowski, 1992; Churchland, Sejnowski, & Arbib, 1992; Smolensky, 1988). For example, one might respond to lines of a particular orientation, or the quality of honesty, or quite possibly something that does not exactly match an established term (Miikkulainen, 1997). Although each neuron responds maximally to a particular microfeature, it responds to a lesser extent to related microfeatures, an organization referred to as *coarse coding* (Hubel & Wiesel, 1965). Not only does a given neuron participate in the encoding of many memories, but each memory is encoded in many neurons. For example, neuron *A* may respond preferentially to sounds of a certain frequency, while its neighbor *B* responds preferentially to sounds of a slightly different frequency, and so forth. However, although *A* responds *maximally* to sounds of one frequency, it responds less reliably to sounds of a similar frequency. The upshot is that an item in memory is *distributed* across a cell assembly that contains many neurons, and likewise, each neuron participates in the storage of many items (Hebb, 1949; Hinton, McClelland, & Rumelhart, 1986). A given experience activates not just *one* neuron, nor *every* neuron to an equal degree; instead, activation is spread across members of an assembly. The same neurons get used and reused in different capacities, a phenomenon referred to as *neural reentrance* (Edelman, 1987).

A final key attribute of memory is the following. Memory is said to be *content addressable,* meaning that there is a systematic relationship between the content of a representation and the neurons where it gets encoded. This emerges naturally as a consequence of the fact that representations activate neurons that are tuned to respond to particular features, so representations that get encoded in overlapping regions of memory share features. As a result, they can thereafter be evoked by stimuli that are similar or "resonant" in some (perhaps context-specific) way (Hebb, 1949; Marr, 1969).

This kind of distributed, content-addressable memory architecture is illustrated schematically in figure 2.1. Each circle represents a microfeature that is maximally responded to by a particular neuron. Circles that are close together respond to microfeatures that are similar or related. The large, diffuse region of whiteness indicates the region of memory activated by the current thought or experience. Note that even if a brain does not possess a neuron that is maximally tuned to a particular microfeature, the brain is still able to encode stimuli in which that microfeature predominates, because representations are distributed across *many* neurons.

The distributed, content-addressable architecture of memory is critically important for creativity. If it were not distributed, there would be no overlap between items that share microfeatures, and thus no means of forging associations between them. If it were not content-addressable, associations would not be meaningful. The upshot of all this is that representations that share features are encoded in overlapping distributions of neurons, and therefore activation can spread from one to another. Thus representations are encoded in memory in a way that takes into account how they are related, *even if this relationship has never been consciously noticed.* We may never have explicitly learned that a white goat is a mammal, but we know it is one nonetheless. It is in this sense that we claimed earlier that people implicitly know more than they have ever explicitly learned. As we will see, this architecture has implications that extend far beyond issues related to the hierarchical structure of knowledge.

It should be pointed out how different this is from a typical computer memory. In a computer memory, each possible input is stored at a unique address. Retrieval is thus a matter of looking at the address in the address register and fetching the item at the specified location. Since there is no *overlap* of representations, there is no means of creatively forging new associations based on newly perceived similarities. The exceptions are computer architectures that are designed to mimic, or are inspired by, the distributed, content-addressable nature of human memory.

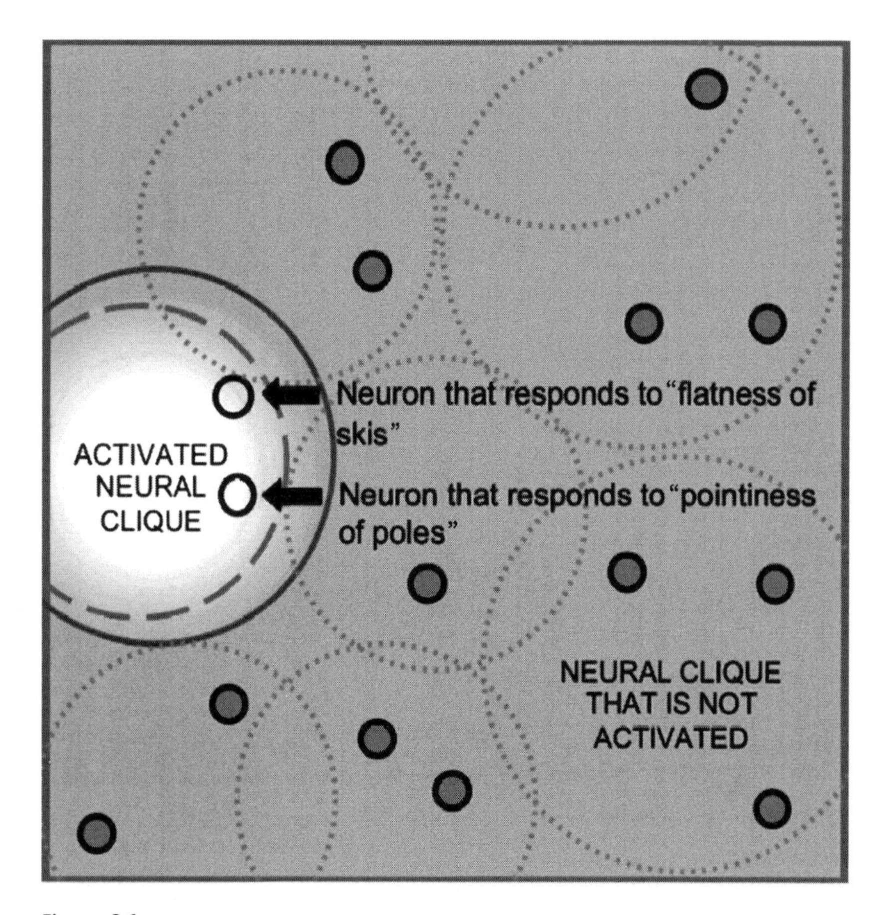

Figure 2.1

Schematized drawing of a portion of a distributed associative memory activated by the thought of snow skiing in an analytic mode of thought. Each small black-ringed circle represents a feature that a particular neuron responds to. The white region indicates the portion of memory that is activated. The activated cell assembly, indicated by the large gray circle, consists of only one neural clique, indicated by the dashed circle. It is composed of neurons that respond to typical features of snow skiing such as the flatness of the skis and the pointiness of the poles. Nonactivated neural cliques are indicated by dotted gray circles.

Forging Unusual Associations through Reconstructive Interference of Memories

A fascinating consequence of the early connectionist findings is that in a distributed, content-addressable memory, not only do representations that share features activate each other, but they can potentially interact in a way that is creative. Even a simple neural network is able to abstract a prototype, fill in missing features of a noisy or incomplete pattern, or create a new pattern on the fly that is more appropriate to the situation than anything that has ever been input to the network (McClelland & Rumelhart, 1986). In fact, similar representations can interfere with one another (Feldman & Ballard, 1982; Hopfield, 1982; Hopfield, Feinstein, & Palmer, 1983). Interestingly, the numerous names for this phenomenon—"crosstalk," "false memories," "spurious memories," "ghosts," and "superposition catastrophe"—are suggestive of a form of thought that, if not outright creative, involves a departure from known reality. Findings from neuroscience are also highly consistent with this phenomenon; indeed, as Edelman (2000) puts it, one does not *retrieve* a stored item from memory so much as *reconstruct* it. That is, an item in memory is never re-experienced in exactly the form it was first experienced, but is colored, however subtly, by what has been experienced in the meantime, and reassembled spontaneously in a way that relates to the task at hand (this is one reason eyewitness accounts cannot always be trusted) (Paterson, Kemp, & Forgas, 2009; Loftus, 1980; Schacter, 2001).

The reconstructive nature of memory, though detrimental in some contexts, is beneficial in others; indeed, we claim it underlies what was referred to earlier as the magic of creativity. Cognitive psychologists have long struggled with the question of how minds generate ideas that are both new and useful. Almost universally they have concluded that creativity must be a process of search not so different from what happens in a typical computer; it must involve sifting through possibilities, perhaps tweaking or exploring them, until an acceptable one is selected (e.g., Newell, Shaw & Simon 1957; Newell & Simon 1972; Finke, Ward, & Smith, 1992; Simonton, 1999). But by approaching creativity from a neuroscientific perspective, and specifically focusing on our chosen level midway between brains and neurons, we see that the content-addressable, reconstructive nature of memory enables the brain to accomplish creative acts *without* recourse to a "search and selection" type explanation.

Because information is encoded in a distributed manner across ensembles of neurons interacting by way of synapses, the meaning of a representation is in part derived from the meanings of other representations that excite similar constellations of neurons; that is why it is sometimes referred to as an *associative memory*. Content addressability ensures that one's brain naturally brings to mind items that are similar to one's current experience in a way that may be unexpected or indefinable yet useful or appealing. Recall that if the regions in memory where two distributed representations are encoded overlap then they share one or more microfeatures. They may have been encoded at different times, under different circumstances, and the correlation between them never explicitly noticed. But the fact that their distributions overlap means that *some* context could come along for which this overlap would be relevant, causing one to evoke the other. There are as many routes by which an association between two representations can be forged as there are microfeatures by which they overlap; that is, there is room for typical as well as atypical connections. Therefore, what gets evoked in a given situation is *relevant*, and that happens *for free*—no search is necessary at all—because memory is content-addressable! The "like attracts like" principle is deeply embedded in our neural architecture.

Moreover, because memory is distributed and subject to crosstalk, if a situation does come along that is relevant to multiple representations, they merge together, a phenomenon that has been termed *reconstructive interference* (Gabora & Saab, 2011). The multiple items may be so similar to each other that you never detect that the recollection is actually a blend of many items. In this case, the distributions of neurons they activate overlap substantially. Or they may differ in mundane ways, as in everyday mindwandering. Alternatively, they may be superficially different but related in a way you never noticed before. In this case, the distributions of neurons they activate overlap with respect to only a few features that in the present context happen to be relevant or important. Alternatively, the present experience may infuse recall of a previous experience that is relevant or important with respect to only a few key features. For example, the person who invented waterskiing may have been sitting on a beach thinking about snow skiing. The SKIING representation merged with the WATER representation, and the idea of WATERSKIING was born. Of course, the invention of waterskiing could have happened differently. The person could have been thinking about Jesus walking on water and seen a boat go by slowly with a fish being pulled in on a fishing line. In this case, waterskiing was born through the merging of the WALK ON WATER representation with the PULLED BY BOAT representation to give MOVE ACROSS WATER PULLED BY BOAT. Note that WATERSKIING (or MOVE ACROSS WATER PULLED BY BOAT) was not waiting in a dormant, predefined state to be selected, nor was it tweaked or mutated in a trial-and-error manner, as implied by accounts of creativity that emphasize search or selection from among multiple concretely defined alternatives (e.g., Newell, Shaw, & Simon, 1957; Simonton, 1999). Reconstructive interference of implicitly present information enables one to "go beyond what one explicitly knows" to solve a problem or express oneself creatively. The greater the extent to which the contributing representations differ, the more likely they are to result in transformative creativity as opposed to mere exploratory creativity.

Resolving Unusual Associations: States of Potentiality and Their Actualization

We saw that reconstructive interference allows us to generate novelty without having to try out lots of possibilities. However, it has a disadvantage. When representations come together for the first time, it is not always clear how they *go* together; indeed, the new idea may barely make sense, as is epitomized by the phrase "half-baked idea." For example, having the idea of waterskiing is a far cry from knowing concretely how one would really ski on water. In the beach scenario, the inventor must figure out that the water-skier is pulled by a boat, and does not need poles. In the Jesus scenario, the inventor must figure out that the water-skier wears skis. In either case, the insight exists initially in a state of potentiality; it is not yet clear how it could actualize. So the *effortful* aspect of creativity involves not generating, testing, and selecting, but *actualizing potential* (Gabora, 2005), by either thinking the idea through or trying it out.

At the moment of insight, different possible realizations of the idea have not yet been conceived. However, they are implicitly present in memory in the sense that each possible realization of the idea activates a different but overlapping constellation of neurons that respond to different sets of microfeatures. For example, one potential realization is that the poles are flattened like paddles. Another potential realization, the most effective one it turns out, involves pulling the skier from a boat.

How Distributed Should a Memory Be for Creative Insight to Take Place?

How much overlap of microfeatures must there be to result in creative insight? At one extreme it could be not distributed at all, like a typical computer memory. If your mind stored each item in just one location as a computer does, then for one experience to remind you of a previous experience, it would have to be *identical* to that previous experience. And since the space of possible experiences is so vast that no two ever *are* exactly identical, this kind of organization would be useless. But at the other extreme, if your memory were *fully distributed*, with each item stored in every location, the crosstalk would be a catastrophe; everything would pretty much remind you of everything else.

The problem of crosstalk is solved by *constraining* the distribution region. One way to do this in neural networks is to use a radial basis function, or RBF (Hancock, Smith, & Phillips, 1991; Holden & Niranjan, 1997; Lu, Sundararajan, & Saratchandran, 1997; Willshaw & Dayan, 1990). Each input activates a hypersphere (sphere with more than three dimensions) of locations, such that activation is maximal at the center k of the RBF and tapers off in all directions according to a (usually) Gaussian distribution of width σ. The result is that one part of the network can be modified without affecting the capacity of other parts to store other patterns. A *spiky activation function* means that σ is small, so that only those locations closest to k get activated, but they are strongly activated. A *flat activation function* means that σ is large, so that locations relatively far from k still get activated, but no location gets strongly activated.

In the brain, the principle of course coding ensures that distributions are constrained. Because neurons respond most reliably to one particular feature and less reliably to similar features, the region of activation falls midway between two extremes—not distributed at all (a one-to-one correspondence between each input and each neuron) and fully distributed (each input activates every neuron). It is because not one neuron, nor every neuron, but a subset of neurons is activated, that one can generate a stream of coherent yet potentially creative thought (Gabora, 2001). The more detail with which items have been encoded in memory, the greater their potential overlap with other items, and the more routes by which one can make sense of the present in terms of the past, adapt old ideas to new situations, or engage in creative thinking.

Insight, Contextual Focus, and "Neurds"

With flat activation, items are evoked in detail, or multiple items with overlapping distributions of microfeatures are evoked simultaneously. Thus, flat activation is conducive to forging remote associations among items not usually thought to be related, or detecting relationships of correlation. Indeed, flat activation would be expected to result in the flat associative hierarchies characteristic of highly creative people (Mednick,

1962). With spiky activation, items are evoked in a compressed form, and few are evoked simultaneously. Thus, it is conducive to mental operations on those items, or deducing relationships of causation. As such, spiky activation would be expected to result in the spiky associative hierarchies characteristic of uncreative people (Mednick, 1962).

One can imagine situations where spiky activation would be useful, as when one needs to stay focused, and accessing remote associations would be distracting. One can also imagine situations where flat activation would be useful, as when conventional problem solving methods do not work well.

It has long been thought that there are two modes of thought, sometimes referred to as *associative* and *analytic* (Freud, 1949; Guilford, 1950; James, 1890), and that we shift along a continuum between these two extremes depending on our situation (Gabora, 2002, 2003). The capacity to shift between the two modes of thought is sometimes referred to as *contextual focus,* because a change from one mode to the other is brought about by the context, through the focusing or defocusing of attention. This is related to *dual process theory,* the idea that cognition employs both implicit and explicit ways of learning and processing information (Chaiken & Trope, 1999; Evans & Frankish, 2009), since analytic thought is believed to involve processing of explicit information whereas associative thought is believed to involve processing of implicit information. Thus, contextual focus entails not just the capacity for both associative and analytic thought, but the capacity to adjust the mode of thought to match the demands of the situation. It seems reasonable that we engage in contextual focus using a mechanism akin to varying the size of the RBF: spontaneously tuning the spikiness of the activation function in response to the situation.

What neural mechanisms might underlie the capacity to shift between associative and analytic modes of thought? It has been shown that the cell assembly involved in the encoding of a particular experience is made up of multiple groups of collectively co-spiking neurons referred to as *neural cliques* (Lin et al., 2005; Lin, Osan, & Tsien, 2006). Techniques that enable their patterns of activation to be mathematically described, directly visualized, and dynamically deciphered reveal that some cliques respond to situation-specific elements of an experience (e.g., where it took place), while others respond to characteristics of varying degrees of generality or abstractness. These range from the type of experience (e.g., being dropped) to characteristics common to many types of experience (e.g., anything dangerous). This has been depicted as a pyramid in which cliques that respond to the most context-specific elements are at the top, and those that respond to the most general elements are at the bottom (Lin et al., 2005).

We can now make a reasonable hypothesis for what is happening at the level of neural cliques during creative thought. Each successive thought activates recruitment of more or fewer neural cliques, depending on the nature of the problem, and how far along one is in solving it. Two well-established phenomena help ensure that a particular thought doesn't recursively reactivate itself. First, if the same neurons are stimulated repeatedly, they become refractory. For the duration of this refractory period they either cannot fire or their response is greatly attenuated. Second, they "team play": a response is produced by a cooperative group of neurons such that when one is refractory another is active. Since the situation-general neurons and the situation-specific neurons are not responding to the same aspects of the situation, they are not entering and leaving their refractory periods in synchrony, making it unlikely that one will think the same, identical thought over and over again (although over a longer time frame one may cycle back to it).

Returning to figure 2.1, we can get a schematic picture of how memory is activated by a particular thought. Recall that the degree to which any given region of memory is activated by the current thought or experience is indicated by the degree of whiteness. The white area thus represents the active cell assembly composed of one or more neural cliques, indicated by dashed gray circles. The further a neuron is from the center of the white region, the less activation it not only *receives* from the current instant of experience but in turn *contributes* to the next instant, and the more likely its contribution is canceled out by that of other simultaneously active locations. Using neural network terminology, we say the broader the region affected, the flatter the activation function, and the narrower the affected region, the spikier the activation function. Figure 2.1 is a schematic portrayal of a portion of the conceptual network of someone sitting on a beach thinking about snow skiing in an analytic mode of thought. The white region is narrow because it is activated in an analytic mode of thought. It includes only neurons that respond to typical features of skiing such as the flatness of the skis and pointiness of the poles.

In a state of defocused attention more aspects of a situation get processed; the set of activated microfeatures is larger, and thus the set of potential associations one could make is larger. Figure 2.2 shows the state of mind of someone sitting at the beach thinking about snow skiing, but here the activation function is flat. Recruitment of neural cliques that respond to abstract elements of the current thought (e.g., sliding across a

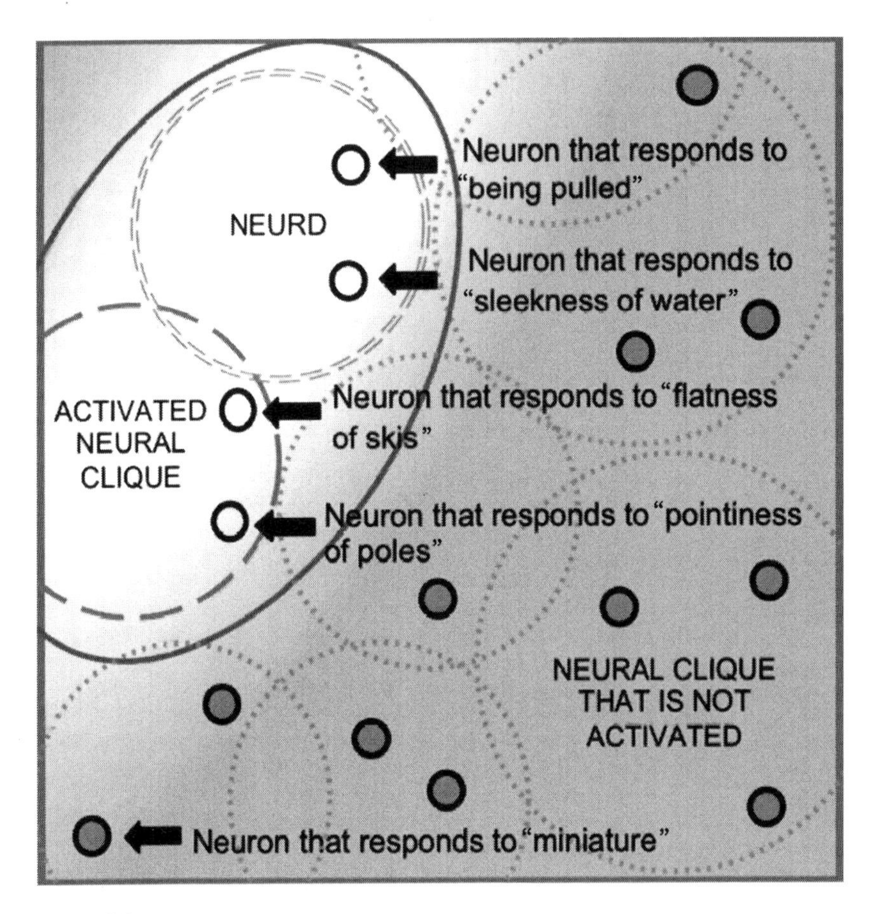

Figure 2.2
In an associative mode of thought, the portion of memory activated by the thought of snow skiing is larger than it was in an analytic mode, as indicated by the size and diffuseness of the white region. The activated cell assembly, indicated by the large oval, now contains more than one neural clique. The initially activated neural clique is indicated by the dashed circle, and the neurd is indicated by the double circle of dashes. The neurd is composed of neurons that respond to features that are not typical of snow skiing, such as "sleekness of water," but which are relevant to the invention of waterskiing. Note that under a different context, such as the task of making skis for a doll, the neurd might have been a different neural clique, containing the neuron that responds to "miniature."

smooth surface) causes the individual to extend the idea of sliding across a surface to the present context of being at the beach. There is reconstructive interference of the skiing concept with this current experience, in which features of skiing (e.g., flatness of skis) are merged with features of water (e.g., that it is liquid, not solid). Some features of water seem irrelevant to skiing, but they are relevant to inventing a means to be able to ski in the summer.

The neural cliques that do not fall within the activated region in figure 2.1 but do fall within the activated region in figure 2.2 are cliques that *would not* be included in a cell assembly if one were in an analytic mode of thought, but *would* be included if one were in an associative mode of thought. We can refer to these neural cliques as *neurds*. Neurds respond to microfeatures that are of marginal relevance to the current thought. Neurds do not reside in any particular portion of memory. The subset of neural cliques that count as neurds is defined by context, and shifts constantly. For each situation, one might encounter a different group of neurds.

The explanation of insight proposed here follows naturally from the above-mentioned discovery of neural cliques that respond to varying degrees of specificity or generality, and from the evidence for contextual focus, as well as the well-established phenomenon that activation of an abstract or general concept causes activation of its instances through a mechanism such as spreading activation (Anderson, 1983; Collins & Loftus, 1975).[1] Given that some neural cliques respond to specific aspects of a situation and others respond to more general or abstract aspects, we have a straightforward mechanism by which contextual focus could be executed. In associative thought, with more aspects of a situation taken into account, more neural cliques are activated, including those responding to specific features, those responding to abstract elements, and those *they* activate through spreading activation. Activation flows from the specific instance to the abstract elements it instantiates, and then to other instances of those abstract elements. The neurds concept thus provides a way of referring to neural cliques that responds to features of these other instances that are not features of the original instance.

It is likely that most of the time, for most individuals, neurds are excluded from activated cell assemblies. Their time to shine comes when one has to break out of a rut. In associative thought, broad activation causes more neural cliques to be recruited, including neurds. This enables the next thought to stray far from the one that preceded it, while retaining a thread of continuity. The associative network is thereby not just penetrated deeply, but traversed quickly, and there is greater potential for representations to overlap in ways they never have before. Thus, the potential to unite previously disparate ideas or concepts is high.

Example and Analysis of an Instance of Insight

We have examined the relationship between contextual focus and the structure of human memory. This synthesis will now be applied to the analysis of a creative act of a sort that is more artistic than the waterskiing example used previously. In keeping with the view that everyone is creative (Beghetto & Kaufman, 2007; Gardner, 1993; Runco, 2004b), the creative act that we analyze is not an earthshaking achievement but a simple event in the life of an everyday person.

The situation that motivates the creative act is the following. Amy, an art student, wants privacy in her bedroom, but the only kind of curtains her landlord can afford are ugly and so thick they would block out all the light and her plants would die. She asks around for secondhand curtains, trying to solve the situation through a straightforward deductive process. Neural cliques that encode memories of various curtains, and other sorts of window coverings such as blinds, are activated. Neurons that respond to attributes of desirable curtains, such as "soft," "hangs (on curtain hooks)," "translucent," "large," and "colorful," are activated.

Her inability to solve the problem rationally eventually leads to a spontaneous and subconscious defocusing of attention. She enters an associative mode of thought, and her activation function becomes flat, such that the associative structure of her memory is more widely probed. Activation of neurons that respond to attributes such as "soft" that are irrelevant to her goals (of obtaining privacy while retaining sufficient light for the plants) decreases, but activation of neurons that respond to attributes such as "translucent" that are relevant to these remains high. Goal-relevant neurds get recruited moving further down Lin et al.'s (2006) feature-encoding pyramid, and her memory begins to respond to not just specific aspects of her situation (e.g., the need for curtains) but to abstract aspects of her situation (e.g., the need for privacy).

Amy starts considering not just different kinds of window coverings such as curtains and blinds, but other attractive ways of covering the window that would allow light to come through. Because memory is content addressable (i.e., there is a systematic relationship between the content of an item and the locations in memory it activates), neural cliques that respond to "translucent" and "colorful," now activated in the context of needing to cover the window, had previously encoded memories of certain acrylic paints that might be opaque enough to provide privacy yet translucent enough to let in light. Activation spreads from neural cliques that respond to "translucent" and "colorful" to neural cliques that respond to other aspects of these acrylic paints, causing her to consciously think of them in this new context, resulting in reconstructive interference of CURTAIN and PAINT to give something like "CURTAIN-PAINT": the idea of using paint to accomplish the job of curtains. It has some attributes of curtains (e.g., window coverage) and some attributes of paint (e.g., hardens upon application), as well as the attributes they both share that allowed the association to be made. Since she is an artist, she could have first thought of the idea of placing a painting in front of the window, realizing that would not let any light in at all, and subsequently realized she could paint directly on the window.[2] Alternatively, it could have been the more general concept WINDOW COVERING that combined with PAINT to give WINDOW PAINT. In either case, the resulting concept combination is new because this particular distributed set of neurons has never been activated together before as an ensemble.

Having hit upon this idea of painting the window, she must determine if it will work in practice. Although in the short run a flat activation function is conducive to creativity, maintaining it would be impractical since the relationship between one thought and the next may be remote; thus a stream of thought lacks continuity. Access to obscure associations might at this point be a distraction. Thus, now she enters a more analytic mode by "decruiting" the neurds, thereby narrowing the region of memory that gets activated. Thought becomes more logical in character because the activation function becomes spikier, thereby affording finer control; fewer locations release their contents to participate in the formation of the next thought. Experimenting with different paints, colors, and brush styles, Amy finds paints that let in light while obscuring visibility.

Once she knows it will work, the actual painting of the window offers more room for creativity, and she returns to a more associative mode of thought. By shifting back and forth along the spectrum from associative to analytic as needed, the fruits of associative thought become ingredients for analytic thought, and vice versa.

Mathematical Description of Creative Insight

We have shown how it is possible to provide a neurobiological account of creative insight. This account is strengthened by the fact that it dovetails with a complementary avenue of research aimed at developing a mathematical account of insight. Since insights involve putting concepts together in new ways or placing them in new contexts, a comprehensive theory of creativity must incorporate a solid theory of *how concepts interact*. However, people use conjunctions and disjunctions of concepts in ways that violate the rules of classical logic; that is, concepts interact in ways that are noncompositional (Barsalou, 1987; Bruza et al., 2009, 2010; Hampton, 1987; Osherson & Smith, 1981). This is true both with respect to properties (e.g., although people do not rate "talks" as a characteristic property of PET or BIRD, they rate it as characteristic of PET BIRD), and exemplar typicalities (e.g., although people do not rate "guppy" as a typical PET, nor a typical FISH, they rate it as a highly typical PET FISH). Because of this, concepts have been resistant to mathematical description.

It has been demonstrated that the non-compositional behavior of concepts can be modeled using a generalization of the formalisms of quantum mechanics (Aerts, 2009; Aerts, Broekaert, Gabora, & Veloz, 2012; Aerts & Gabora, 2005a,b; Gabora & Aerts, 2002, 2009; Gabora, Rosch, & Aerts, 2008; Gabora, Scott, & Kauffman, in press). The state $|\psi\rangle$ of an entity is written as a linear superposition of a set of basis states $\{|\phi_i\rangle\}$ of a complex Hilbert space $\mathcal{H}$. Hence $|\psi\rangle = \Sigma_i c_i|\phi_i\rangle$ where each complex number coefficient c_i of the linear superposition represents the contribution of each component state $|\phi_i\rangle$ to the state $|\psi\rangle$ of the entity; more specifically, the square of the absolute value of each coefficient equals the weight of the corresponding component basis state with respect to the global state. The choice of basis states is determined by the observable to be measured, and the basis states corresponding to this observable are called *eigenstates* of the observable. Upon measurement, the state of the entity *collapses* to one of the eigenstates.

In the quantum-inspired state context property (SCOP) theory of concepts, the basis states represent instances or exemplars of a concept, and the measurement is the context that causes a particular state to be evoked. A concept is defined in terms of (i) its set of states or exemplars Σ, each of which consists of a set $\mathcal{L}$ of relevant properties; (ii) set $\mathcal{M}$ of contexts in which it may be relevant; (3) a function ν that describes the applicability or *weight* of a certain property for a specific state and context; and (4) a function μ that describes the transition probability from one state to another under the influence of a particular context.

Let us now see how the example would be described using the formalism. The concept WINDOW COVERING consists of the set Σ of states such as CURTAIN, BLINDS, and so forth. For simplicity, let us say that one's initial conception of WINDOW COVERING, represented by vector $|w\rangle$ of length equal to 1, is a superposition of only two possibilities. The possibility that it is a blind is denoted by unit vector $|b\rangle$. The possibility that it is a curtain is denoted by unit vector $|c\rangle$. This can be described by the equation $|w\rangle = d_0|b\rangle + d_1|c\rangle$, where d_0 and d_1 are the default amplitudes of $|b\rangle$ and $|c\rangle$, respectively. States are always represented by unit vectors, and all vectors of a decomposition such as $|b\rangle$ and $|c\rangle$ have unit length, are mutually orthogonal, and generate the whole vector space; thus, $|d_0|^2 + |d_1|^2 = 1$.

Amy realizes that she must constrain the window covering to be inexpensive, and this may be challenging, so she shifts to a more associative mode of thought. The context "inexpensive" is denoted *i*. Amy knows that blinds are expensive. Therefore, i_0, the new amplitude of $|b_i\rangle$, equals 0. Since Amy does not know of any inexpensive curtains, i_1 is close to 0. Activation of the set L of desirable properties of WINDOW COVERING, for example, "translucent" but not "transparent," spreads to other concepts in the individual's associative network for which these properties are relevant. Concepts that share these properties with CURTAIN and BLINDS become candidate members of the set Σ of relevant states of WINDOW COVERING. It becomes apparent that the concept WINDOW COVERING in the context "inexpensive," denoted $|w_i\rangle$, has a third term, such that $|w_i\rangle = i_0|b_i\rangle + i_1|c_i\rangle + i_2|p_ic_i\rangle$, where $|p_ic_i\rangle$ represents the possibility that PAINT functions as a WINDOW COVERING.

Since PAINT makes an inexpensive WINDOW COVERING, $|i_3|$ is large. Therefore, in the context "inexpensive," the concept WINDOW COVERING has a high probability of collapsing to WINDOW PAINT, which (for reasons we do not have space to delve into here) there is reason to believe is most aptly described as an entangled state of the concepts WINDOW COVERING and PAINT. Entanglement introduces interference of a quantum nature, and hence the amplitudes are complex numbers (Aerts, 2009). If this collapse takes place, WINDOW PAINT is thereafter a new state of both concepts WINDOW COVERING and PAINT.

This example shows that the above neurobiological account of insight, based on context-dependent overlap of properties, is amenable to formal mathematical modeling. It also shows that concept combination always entails, at least to some extent, both exploration and transformation of conceptual spaces. CURTAIN combined with PAINT in the context "inexpensive" when the inventor of CURTAIN PAINT explored the space of possible uses of a worn-out CURTAIN. Moreover, when CURTAIN combined with PAINT in the context "inexpensive," the conceptual space associated with items that could be considered inexpensive was altered to include a new kind of PAINT. Note how in the quantum representation, probability is treated as arising not from a lack of information *per se*, but from the limitations of any particular context (even a "default" context). Though we do not have space to discuss them here, mathematical (Veloz, Gabora, Eyjolfson, & Aerts, 2011) as well as computational (DiPaola & Gabora, 2009; Gabora, Chia, & Firouzi, 2013; Gabora & DiPaola, 2012) models have also been developed of the capacity to shift between convergent/analytic and divergent/associative modes of thought.

Summary and Conclusions

This chapter provides a tentative but sound and plausible explanation of how the creative process works at the level of distributed ensembles of neurons. We believe that it is at this level that one can gain insight into what we call the magic of creativity: the ability to pull out of memory something new and appropriate that was never explicitly stored there.

The brain is able to do this because of the ingenious way that representations are laid down in memory. First, representations are *distributed* across assemblies of neurons, sometimes referred to as "neural cliques." Each neuron is "tuned" to respond maximally to a particular microfeature: a particular orientation of a line, for example, or something more abstract. It responds less reliably to lines with nearby orientations. This leads to another basic principle of memory, *coarse coding*: each neuron participates in the encoding of many experiences. Thus, a memory of a particular event involves activation of not one neuron but a constellation of them. Memory is also *content addressable,* meaning there is a relationship between the content of an item and which neurons respond to it. Thus, similar events activate overlapping constellations of neurons. This means that items are encoded in memory not just in terms of their features, but in terms of how they relate to each other; the associative structure of the brain reflects underlying statistical regularities in ones' experiences. It is this implicit knowledge of how things are related, indeed related in ways one may never have consciously noticed, that is called upon in the creative process.

Because of the distributed, content-addressable architecture of memory, multiple items may be evoked simultaneously and can merge to give rise to a thought that bears some similarity to these multiple items, but which is identical to none of them. The distributions of neurons they activate might overlap substantially, or they might overlap with respect to only a few features. Because of the phenomenon of *reconstructive interference,* the result may be an insight that combines elements of both. The greater the extent to which they differ, the greater the extent to which the insight will appear to be an instance of transformative rather than merely exploratory creativity.

When one is stumped and in need of a creative solution, or in a situation conducive to creative self-expression, one defocuses attention and enters a more associative form of thought, through a process that is believed to work in a manner similar to flattening the activation function in a neural network. Associative thought causes activation of not just neural cliques that respond to situation-specific aspects of an experience, but neural cliques that respond to general or abstract aspects. Those neural cliques that respond to atypical aspects of the situation, and thus are activated in associative but not analytic thought, are referred to as *neurds*. Items encoded previously to neurds are superficially different from the present situation yet share aspects of its deep structure. Therefore, the recruitment of neurds increases the probability of forging associations that are seemingly irrelevant yet vital to the creative task. By responding to abstract or atypical features of the situation, neurds effectively draw remote associates into the conceptualization of the task. If an insightful association is made, one may enter a more analytic or convergent mode of thought through a process akin to increasing the spikiness of the activation function, by deactivating neurds. Analytic thought discourages potentially disruptive associations and is thus conducive to simply getting a job done.

It is important to stress that the neurobiological interpretation of associative thought given here does not entail making all associated items within a given semantic distance more accessible (like a diverging beam of light), but only those that are contextually relevant. Moreover, it does not necessarily entail evoking *multiple* items. Associative thought may evoke just a single item, something that had never been explicitly stored in memory, which merges multiple previously encoded items together.

It is interesting to consider the long-term consequences of the proclivity to shift readily into a state of defocused attention. More features of any given experience evoke "ingredients" from memory for the next experience. Some aspects of the external world get ignored because one is busy processing previous material, but if something does manage to attract attention, it tends to be considered from multiple angles before one settles into a particular interpretation of it. The end result is that one's understanding of the world becomes increasingly unique, and this uniqueness may be reflected in one's creative output.

The explanation of creativity put forward here is incomplete, particularly with respect to the role of motivation and emotions. Moreover, a complete neuroscientific account of creativity will include an explanation of how events at the level of distributed cell assemblies dovetail with, on the one hand, activity in particular areas of the brain, and on the other hand, the intracellular events that make the formation of new representations possible. We believe, however, that some of the more mysterious aspects of creativity have now been solved.

Acknowledgments

This research was funded by grants from the Sciences and Engineering Research Council of Canada (NSERC) and the Fund for Scientific Research, Flemish Government of Belgium.

Notes

1. Thus, for example, based on a set of free-association norms data collected from 6,000 participants using over 5,000 words, the probability that, given the word PLANET, the first word that comes to mind is EARTH is 0.61, and the probability that it is MARS is 0.10 (Nelson, McEvoy, & Schreiber, 2004). Note that there is some empirical support for an alternative to spreading activation as an explanation for this kind of association data, referred to as "spooky activation at a distance" (Nelson, McEvoy & Pointer, 2003).

2. This possibility for how the idea came about was suggested by Richard Gabriel.

References

Aerts, D. (2009). Quantum structure in cognition. *Journal of Mathematical Psychology, 53*, 314–348.

Aerts, D., Broekaert, J., Gabora, L., & Veloz, T. (2012). The guppy effect as interference. In *Proceedings of the Sixth International Symposium on Quantum Interaction.* June 27–29, Paris.

Aerts, D., & Gabora, L. (2005a). A state-context-property model of concepts and their combinations I: The structure of the sets of contexts and properties. *Kybernetes, 34*, 151–175.

Aerts, D., & Gabora, L. (2005b). A state-context-property model of concepts and their combinations II: A Hilbert space representation. *Kybernetes, 34*, 176–205.

Anderson, J. R. (1983). A spreading activation theory of memory. *Journal of Verbal Learning and Verbal Behavior, 22*, 261–295.

Andrews-Hanna, J. R., Reilder, J. S., Huang, C., & Buckner, R. L. (2010). Evidence for the default network's role in spontaneous cognition. *Journal of Neurophysiology, 104*, 322–335.

Barsalou, L. (1987). The instability of graded structure: Implications for the nature of concepts. In U. Neisser (Ed.), *Concepts and conceptual development: Ecological and intellectual factors in categorization*. Cambridge: Cambridge University Press.

Beghetto, R. A. & Kaufman, J. C. (2007). Toward a broader conception of creativity: A case for "mini-c" creativity. *Journal of Aesthetics, Creativity, and the Arts, 1*, 73–79.

Blasdel, G. G., & Salama, G. (1986). Voltage sensitive dyes reveal a modular organization in monkey striate cortex. *Nature, 321*, 579–585.

Boden, M. A. (1990). *The creative mind: Myths and mechanism*. London: Abacus.

Bruza, P., Kitto, K., Nelson, D., & McEvoy, C. (2009). Is there something quantum-like in the human mental lexicon? *Journal of Mathematical Psychology, 53*, 362–377.

Bruza, P. D., Kitto, K., Ramm, B., Sitbon, L., Blomberg, S., & Song, D. (2010). Quantum-like non-separability of concept combinations, emergent associates and abduction. *Logic Journal of the IGPL, 20*, 445–457.

Chaiken, S., & Trope, Y. (1999). *Dual-process theories in social psychology*. New York: Guilford Press.

Chandrasekharan, S. (2009). Building to discover: A common coding model. *Cognitive Science, 33*, 1059–1086.

Churchland, P. S., & Sejnowski, T. (1992). *The computational brain*. Cambridge, MA: MIT Press.

Churchland, P. S., Sejnowski, T. J., & Arbib, M. A. (1992). The computational brain. *Science, 258*, 1671–1672.

Collins, A. M., & Loftus, E. F. (1975). A spreading-activation theory of semantic processing. *Psychological Review, 82*, 407–428.

Dayan, P., & Abbott, L. F. (2001). *Theoretical neuroscience: Computational and mathematical modeling of neural systems*. Cambridge, MA: MIT Press.

DiPaola, S., & Gabora, L. (2009). Incorporating characteristics of human creativity into an evolutionary art algorithm. *Genetic Programming and Evolvable Machines, 10*(2), 97–110.

Edelman, G. (1987). *Neural Darwinism: The theory of neuronal group selection*. New York: Basic Books.

Edelman, G. (2000). *Bright air, brilliant fire: On the matter of the mind*. New York: Basic Books.

Eliasmith, C., & Anderson, C. H. (2003). *Neural engineering: Computation, representation, and dynamics in neurobiological systems*. Cambridge, MA: MIT Press.

Evans, J., & Frankish, K. (Eds.). (2009). *In two minds: Dual processes and beyond*. New York: Oxford University Press.

Feldman, J., & Ballard, D. (1982). Connectionist models and their properties. *Cognitive Science, 6*, 204–254.

Finke, R. A., Ward, T. B., & Smith, S. M. (1992). *Creative cognition: Theory, research, and applications*. Cambridge, MA: MIT Press.

Freud, S. (1949). *An outline of psychoanalysis*. New York: Norton.

Gabora, L. (2001). *Cognitive mechanisms underlying the origin and evolution of culture*. Ph.D. Dissertation, Free University of Brussels.

Gabora, L. (2000). Toward a theory of creative inklings. In R. Ascott (Ed.), *Art, technology, and consciousness* (pp. 159–164). Bristol, UK: Intellect Press.

Gabora, L. (2002). Cognitive mechanisms underlying the creative process. In T. Hewett & T. Kavanagh (Eds.), *Proceedings of the Fourth International Conference on Creativity and Cognition*, 126–133. October 13–16, Loughborough University, UK. New York: ACM Press.

Gabora, L. (2003). Contextual focus: A cognitive explanation for the cultural transition of the Middle/Upper Paleolithic. In R. Alterman & D. Kirsh (Eds.), *Proceedings of the 25th Annual Meeting of the Cognitive Science Society*. July 31–August 2, Boston, MA. Hillsdale, NJ: Lawrence Erlbaum.

Gabora, L. (2005). Creative thought as a non-Darwinian evolutionary process. *Journal of Creative Behavior, 39*, 262–283.

Gabora, L. (2010). Revenge of the "neurds": Characterizing creative thought in terms of the structure and dynamics of human memory. *Creativity Research Journal, 22*, 1–13.

Gabora, L. (under revision). The honing theory of creativity.

Gabora, L., & Aerts, D. (2002). Contextualizing concepts using a mathematical generalization of the quantum formalism. *Journal of Experimental & Theoretical Artificial Intelligence, 14*, 327–358.

Gabora, L., & Aerts, D. (2009). A mathematical model of the emergence of an integrated worldview. *Journal of Mathematical Psychology, 53*, 434–451.

Gabora, L., Chia, L., & Firouzi, H. (2013). A computational model of two cognitive transitions underlying cultural evolution. In *Proceedings of the 35th Annual Meeting of the Cognitive Science Society*. Austin, TX: Cognitive Science Society.

Gabora, L., & DiPaola, S. (2012). How did humans become so creative? In *Proceedings of the International Conference on Computational Creativity*. May 31–June 1, 2012, Dublin, Ireland.

Gabora, L., Rosch, E., & Aerts, D. (2008). Toward an ecological theory of concepts. *Ecological Psychology, 20,* 84–116.

Gabora, L., & Saab, A. (2011). Creative interference and states of potentiality in analogy problem solving. In *Proceedings of the Annual Meeting of the Cognitive Science Society*. July 20–23, 2011, Boston MA.

Gabora, L. Scott, E., & Kauffman, S. (in press). A quantum model of exaptation: Incorporating potentiality into biological theory. *Progress in Biophysics & Molecular Biology*.

Gardner, H. (1993). *Frames of mind: The theory of multiple intelligences*. New York: Basic Books.

Guilford, J. P. (1950). Creativity. *American Psychologist, 5,* 444–454.

Hampton, J. (1987). Inheritance of attributes in natural concept conjunctions. *Memory & Cognition, 15,* 55–71.

Hancock, P. J. B., Smith, L. S., & Phillips, W. A. (1991). A biologically supported error-correcting learning rule. *Neural Computation, 3,* 201–212.

Hebb, D. (1949). *The organization of behavior*. New York: Wiley.

Hebb, D. (1980). *Essay on mind*. Hillsdale, NJ: Lawrence Erlbaum.

Hinton, G., McClelland, J., & Rumelhart, D. (1986). Distributed representations. In D. Rumelhart, J. McClelland, & the PDP Research Group (Eds.), *Parallel distributed processing* (Vol. 1, pp. 77–109). Cambridge, MA: MIT Press.

Holden, S. B., & Niranjan, M. (1997). Average-case learning curves for radial basis function networks. *Neural Computation, 9,* 441–460.

Hopfield, J. J. (1982). Neural networks and physical systems with emergent collective computational abilities. *Proceedings of the National Academy of Sciences (Biophysics), 79,* 2554–2558.

Hopfield, J. J., Feinstein, D. I., & Palmer, R. D. (1983). "Unlearning" has a stabilizing effect in collective memories. *Nature, 304,* 159–160.

Hubel, D. A., & Wiesel, T. N. (1965). Receptive fields and functional architecture in two non-striate visual areas (18 and 19) of the cat. *Journal of Neurophysiology, 28,* 229–289.

James, W. [1890] (1950). *The principles of psychology*. New York: Dover.

Lin, L., Osan, R., Shoham, S., Jin, W., Zuo, W., & Tsien, J. Z. (2005). Identification of network-level coding units for real-time representation of episodic experiences in the hippocampus. *Proceedings of the National Academy of Sciences of the United States of America, 102,* 6125–6130.

Lin, L., Osan, R., & Tsien, J. Z. (2006). Organizing principles of real-time memory encoding: Neural clique assemblies and universal neural codes. *Trends in Neurosciences, 29*, 48–57.

Loftus, E. F. (1980). *Memory: Surprising new insights into how we remember and why we forget.* Reading, MA: Addison-Wesley.

Lu, Y. W., Sundararajan, N., & Saratchandran, P. (1997). A sequential learning scheme for function approximation using minimal radial basis function neural networks. *Neural Computation, 9*, 461–478.

Marr, D. A. (1969). Theory of the cerebellar cortex. *Journal of Physiology, 202*, 437–470.

McClelland, J. L., & Rumelhart, D. E. (1986). A distributed model of memory. In D. Rumelhart, J. McClelland, & the PDP Research Group (Eds.), *Parallel distributed processing: Explorations in the microstructure of cognition* (Vol. II, Ch. 17). Cambridge, MA: MIT Press.

Mednick, S. (1962). The associative basis of the creative process. *Psychological Review, 69*, 220–232.

Miikkulainen, R. (1997). Natural language processing with subsymbolic neural networks. In A. Brown (Ed.), *Neural network perspectives on cognition and adaptive robotics.* Bristol, UK: Institute of Physics Press.

Nelson, D. L., McEvoy, C. L., & Pointer, L. (2003). Spreading activation or spooky activation at a distance? *Journal of Experimental Psychology: Learning, Memory, and Cognition, 29*, 42–52.

Nelson, D. L., McEvoy, C. L., & Schreiber, T. A. (2004). The University of South Florida word association, rhyme and word fragment norms. *Behavior Research Methods, Instruments, & Computers, 36*, 408–420.

Newell, A., & Simon, H. (1972). *Human problem solving.* Edgewood Cliffs, NJ: Prentice-Hall.

Newell, A., Shaw, C., & Simon, H. (1957). The process of creative thinking. In H. E. Gruber, G. Terrell, & M. Wertheimer (Eds.), *Contemporary approaches to creative thinking* (pp. 153–189). New York: Pergamon.

Ohlsson, S. (1992). Information-processing explanations of insight and related phenomena. In M. T. Keane & K. J. Gilhooly (Eds.), *Advances in the psychology of thinking* (Vol. 1, pp. 1–44). New York: Harvester Wheatsheaf.

Osherson, D., & Smith, E. (1981). On the adequacy of prototype theory as a theory of concepts. *Cognition, 9*, 35–58.

Paterson, H. M., Kemp, R. I., & Forgas, J. P. (2009). Co-witnesses, confederates, and conformity: The effects of discussion and delay on eyewitness memory. *Psychiatry, Psychology, and Law, 16*, S112–S124.

Runco, M. A. (2004a). Creativity. *Annual Review of Psychology, 55*, 657–687.

Runco, M. A. (2004b). Everyone is creative. In R. J. Sternberg, E. L. Grigorenko, & J. L. Singer (Eds.), *Creativity: From potential to realization* (pp. 21–30). Washington, DC: American Psychological Association.

Schacter, D. L. (2001). *The seven sins of memory: How the mind forgets and remembers*. Reading, MA: Houghton Mifflin.

Signer, J. L., & Antrobus, J. S. (1963). A factor-analytic study of day-dreaming and conceptually related cognitive and personality variables. *Perceptual and Motor Skills, 17*, 187–209.

Simonton, D. K. (1999). Creativity as blind variation and selective retention: Is the creative process Darwinian? *Psychological Inquiry, 10*, 309–328.

Smallwood, J., & Schooler, J. W. (2006). The restless mind. *Psychological Bulletin, 132*, 946–958.

Smith, E. E., & Kosslyn, S. M. (2007). *Cognitive psychology: Mind and brain*. Upper Saddle River, NJ: Pearson Prentice Hall.

Smolensky, P. (1988). On the proper treatment of connectionism. *Behavioral and Brain Sciences, 11*, 1–43.

Veloz, T., Gabora, L., Eyjolfson, M., & Aerts, D. (2011). A model of the shifting relationship between concepts and contexts in different modes of thought. In *Proceedings of the Fifth International Symposium on Quantum Interaction*. June 27–29, 2011, Aberdeen, UK.

Willshaw, D., & Dayan, P. (1990). Optimal plasticity from matrix memories: What goes up must come down. *Neural Computation, 2*(1), 85–93.

3 There Is Room for Conditioning in the Creative Process: Associative Learning and the Control of Behavioral Variability

W. David Stahlman, Kenneth J. Leising, Dennis Garlick, and Aaron P. Blaisdell

J. S. Bach created impressive and beautiful canons and fugues through creative manipulations of musical scales. Archimedes invented new methods of geometry to determine that the exact value of pi lay between two fractions: $3^{10/71}$ and $3^{1/7}$. Organic chemist Friedrich August Kekulé von Stradonitz had a dream about a snake biting its own tail, an analogy that provided him with the insight that important molecular compounds have a ring structure. Einstein discovered the fundamental relationship between time, space, matter, and energy. Mandelbrot observed that complex structures of real-world objects (mountains, coastlines, snowflakes, etc.) conform to hierarchies of fractal patterns repeated at multiple scales of observation. Although great scientists, mathematicians, and artists are revered for their creative genius, ordinary people in their daily lives also perform creative acts. When a student has trouble understanding a difficult concept, the teacher may apply a simple analogy to relate the concept to something familiar to the student, thereby engendering new understanding. If you take off your shoe to pound a nail into the wall because you've misplaced your hammer, you have engaged in a creative act.

Yet, creativity is not the sole domain of Mankind, of da Vinci, of Beethoven, of Shakespeare and Jimi Hendrix. Although a humble pigeon, possessing a brain the size of a cashew, cannot write beautiful sonnets, it and other nonhuman animals are equipped with the psychological processes that contribute to at least nominally creative behavior. Animals face problems and challenges in finding food and mates, avoiding predators, and thwarting competitors; creative behavior may play an adaptive role in their survival. What are the defining characteristics of creative behavior, and how can it be studied in humans and animals alike? Creativity can be defined as the tendency to generate new ideas or behaviors that may be useful in solving problems. Thus, a creative act is defined by both its novelty and its value in some context. We present a view of creativity from the perspective of animal researchers studying behavioral processes fundamental to the creative act. Our position is based on several principles on the nature of creative action.

(1) Creativity is dependent on mechanisms related to the production of novel and variable behavior. Without mechanisms for novel action, creative behavior is impossible, as a creative act is by definition an action that is, on some level, new.
(2) A creative act is not mere noise, but serves a useful end. That is to say, the production of novel and variable behavior is not in *itself* "creativity," but it is critically important to the creative act.
(3) Processes that generate novelty and variability in behavior can be accessed and controlled in associative conditioning procedures. The distribution of an animal's behavior is controlled by prior learning about the predictive value of stimuli (as in Pavlovian conditioning) and the consequences of its actions (as in instrumental conditioning).

The first two principles have widespread popularity among both the lay population and creativity researchers. A behavior must be both novel and (at least potentially) adaptive in order to qualify as creative. We do not take these to be controversial statements, nor do we believe them to be anything other than obvious. However, the associative learning mechanisms that produce variability in behavior have received insufficient attention in the literature of creativity. Therefore, in this chapter, we will focus primarily on the role of conditioning as it pertains to novel and variable action (i.e., Principle 3). This is a relatively narrow focus on the topic of creativity, but we believe it to be extraordinarily important.[1]

Creativity researcher Margaret A. Boden (2004) wrote, "Chance with judgment can give us creativity; chance alone, certainly not" (p. 237). This statement is not to imply that chance is unimportant. Indeed, it is necessary (though insufficient) for creativity. It is important to note that this chapter deals more with the processes that produce novel behavior (i.e., "chance alone"), and is not meant to provide a complete account of the creative act in nonhuman animals. This chapter seeks to examine how the behavioral grist for the creativity mill is generated in the first place. We will seek to answer the question of what processes control the emergence of new simple behaviors.

Spontaneous Behavior and the Problem of Novelty

Classically, a behavior may be described as instrumental if it is *emitted* by an organism, rather than *elicited* by the stimulus circumstances within which the organism finds itself. A behavior that is under a great deal of control may properly be described as an instrumental (a.k.a. operant) action; behavior that emerges relatively uncontrolled from the animal's context is something else (e.g., reflex, Pavlovian conditioned response). For example, kicking a soccer ball toward your friend as play behavior is an instrumental action, directly under your control; on the other hand, kicking your leg upward when your doctor raps your patellar tendon is a simple, uncontrolled reflexive response to a stimulus.

An interesting aspect of spontaneous behavior, and one that is often overlooked, is the fact that it is never exactly the same. Certainly, there are actions that differ only a very small amount from other, very similar actions that have been performed in the past. For example, the behavior of tying one's shoes is unlikely to be *very* different from its prior instances. However, there is little doubt that on the level of the organism, there must be minute differences in examples of even this well-trained action: the eyes of the person may flutter in a new way; the position of the foot relative to the hands is likely to be different; the laces themselves subtly degrade with each passing day and with each successive knot, engendering slightly less resistance to the tying process; and so on. Even well-trained athletes show subtle variation in the execution of a well-practiced skill (Bartlett, Wheat, & Robins, 2007). There is a great deal of scientific evidence that behavior commonly thought to be invariant in fact has small amounts of variation from instance to instance (e.g., Bartlett et al., 2007; Brainard & Doupe, 2001).

A classic behaviorist view has a difficult time accounting for this truth (Epstein, 1991). Indeed, that current behavior is never precisely the same as prior action is a potential concern for a psychological position such as B. F. Skinner's brand of radical behaviorism. Radical behaviorism rests on the notion that the genetic makeup and environmental history (i.e., reinforcement schedules) of the organism are fundamentally critical for the production of behavior. However, a piece of the puzzle is missing: where does a new behavior come from in the first place?

This is problematic. Speaking in the behaviorist tradition, an animal's actions are selected by the consequences of those actions; should a response (e.g., lever-pressing) be followed by a favorable circumstance (e.g., food delivery), then the response will be more likely to occur in the future. At its heart this is a redescription of Thorndike's (1927) law of effect, which remains a fundamental principle of instrumental learning (see Dennett, 1975). However, it is important to note that this kind of account *assumes the response* before learning occurs. This account can describe the selection of a particular action as becoming relatively stronger or relatively weaker, but it does so by accepting that the response is already part of the individual's repertoire, ready to be selected by the environment. The law of effect is silent on the origins of *new* behavior; one or more of an animal's many expressed behaviors may be selected and strengthened. Skinner himself reflected a lack of concern for the processes that generate new behavior, attributing novel creative action to "chance" and random "mutations" (Skinner, 1970). Furthermore, Skinner (1966) argued that the process by which an animal moves toward a solution to a problem "does not ... necessarily reflect an important behavioral process" (p. 240). Epstein (1991) reflects on this curious dismissal of processes underlying novel action:

> Skinner took generativity for granted, relying on broad-brush explanations of creativity or on no explanations at all—even suggesting that the creative process was "not important." This fits his two-factor form of determinism. Nontrivial mechanisms of variation might have made the organism seem a little too autonomous for Skinner's liking. (p. 365)

Surely the environment can select action, but clearly there must be an account for the emergence of new action (Epstein, 1990). As Epstein (1991) put it, "Skinner's deterministic dyad always needed another factor: Behavior is determined by genes, environmental history, and certain *mechanisms of variability*" (p. 363). It is clear that we must now discuss some of the work that represents the recent study of the production of novel and variable behavior in animals. We devote the next sections to these mechanisms of variability generation and control.

Explicitly Reinforced Variability in Behavior

Many early studies indicated that variability could be modulated by reinforcement; it is uncontroversial to acknowledge that the schedule of rewarding outcomes has long been known to predictably produce differential levels of behavioral variability in animals. In a bar-pressing task, Schoenfeld, Harris, and Farmer (1966) reinforced rats when their successive interresponse times fell into two distinctive temporal class intervals. This restriction on rewarded actions resulted in a very low level of variation in the rats' bar-pressing behavior. Bryant and Church (1974) reinforced rats for performing on a pair of levers; if reinforcement was contingent on the rats alternating their responses (e.g., left, then right) on only 50 percent of trials, rats tended to develop relatively stereotyped behavior to a single lever. If reinforcement was contingent on the rats alternating levers on 75 percent of trials, however, the rats' bar-pressing behavior became so variable as to be indistinguishable from random. Blough (1966) conditioned pigeons to peck at a keylight for food, with reinforcement only being delivered immediately following unusual interresponse intervals. This training resulted in the pigeons responding to the keylight with interresponse times approaching a random (i.e., highly variable) distribution. Pryor, Haag, and O'Reilly (1969) performed a study where they specifically reinforced novel behaviors (e.g., jumps, flips) in porpoises; the porpoises responded to this training by producing behavior that had never been seen before. In this case, reinforcement of novelty, of uncommon behavior, increased behavioral variation in the porpoises. Indeed, their experiment concluded at the point where classification of the porpoises' increasingly variable behavior became virtually impossible.

Page and Neuringer (1985) extended the evidence that suggests that variation is an instrumentally controlled component of behavior. Page and Neuringer were careful to point out that there are two possible sources of behavioral variation: (1) incidental variation due to the schedule of reinforcement (e.g., Schwartz, 1982), and (2) direct reinforcement of variation itself. This latter possibility is the suggestion that reinforcing an animal for behaving variably will, in and of itself, engender higher levels of variability in behavior (e.g., Pryor et al., 1969). In one experiment, Page and Neuringer directly investigated whether variability is a reinforceable dimension of behavior by comparing pigeons' response variation on a Lag 50 schedule (i.e., a response sequence was only reinforced if it was different from each of the last 50 sequences) to responses by a control group of pigeons that received the same rate of reinforcement, but where delivery was not contingent on the novelty of their response sequences. They found that variation was significantly greater in pigeons under the experimental Lag 50 schedule than the yoked control procedure, clearly demonstrating that variation can be directly manipulated through reinforcement. These findings also suggest that data from previous studies (e.g., Blough, 1966; Bryant & Church, 1974) were not necessarily by-products of the experimental reinforcement schedules, but instead may have been due to specific reinforcement of variability itself. In another experiment, Page and Neuringer (1985) demonstrated that the reinforced variability of responding could be brought under stimulus control in a manner similar to other aspects of behavior (e.g., response force, rate of responding; see also Ross & Neuringer, 2002; Denney & Neuringer, 1998; Neuringer, 1993; Morgan & Neuringer, 1990). This evidence illustrates that variability can clearly be an operant (Neuringer, 2002)—that is to say, the novelty of the form of a response can be increased with reinforcement of novel performance itself.

There are two potential explanations regarding the underlying mechanism that animals use to produce variation (Neuringer, 2004). One possible explanation is couched in terms of retrospective memory, or memory of previous responses. Animals that can remember recent behaviors can learn to avoid repeating these actions if high operant variability is being reinforced. Strong evidence for this kind of explanation comes from studies of rats performing in a radial arm maze; in a standard eight-arm maze, rats quickly learn to run down each of the arms without repetitions in order to efficiently obtain food placed at the end of each arm. This behavioral pattern is clearly efficient, as revisits to arms are not rewarded. Memory of previously visited arms prevents revisits and enables requisite variation to maximize the efficiency of retrieving food rewards in the radial maze. The evidence suggests that many animals are capable of using retrospective memory processes to generate enough variability so as to navigate the maze and collect food efficiently (Cook, Brown, & Riley, 1985; see also Kesner & Despain, 1988).

A second mechanism that may account for operant variability is a stochastic (i.e., random) behavior-generation process (Brembs, 2011). Evidence has shown that when variability is explicitly reinforced it tends to approach a random distribution (e.g., Blough, 1966; Neuringer, 1986; Page & Neuringer, 1985). Pharmacological manipulation of memory in rats also supports the stochastic generator hypothesis of behavior. On the one hand, ethanol disrupts short-term memory processes; presumably, failure of performance on a task that requires repetitive behavior is due to a failure in remembering the response sequence and/or recent behavior. The administration of ethanol has a marked deleterious effect on performance in rats rewarded for a stereotyped sequence of lever presses (e.g., left-left-right-right; McElroy & Neuringer, 1990). On the other hand, injections of ethanol have virtually no effect on operant performance when rats are reinforced for high variability in behavior (Cohen, Neuringer, & Rhodes, 1990; McElroy & Neuringer, 1990). Memory therefore seems critical for operant repetition, but not for operant variability (Neuringer, 1991). The results of Page and Neuringer (1985) represent further evidence against the memory account of behavioral variation. Remember, in this study a group of pigeons' response sequences were reinforced on a Lag 50 contingency. Although the birds quickly learned to respond variably under this schedule, it seems highly doubtful that they were able to remember and avoid each of their previous fifty response sequences (Neuringer, 2004; but see Cook, Levison, Gillett, & Blaisdell, 2005, for evidence of the prodigious capacity of pigeon memory). Another study by Neuringer (1991) directly compared memory and random processes in rats. Neuringer inserted retention intervals (RIs) of varying length between individual responses across two levers; response sequences in one group were reinforced on a Lag 5 schedule (e.g., Group VAR). Whereas these RIs should interfere with memory for prior responses, they should have no deleterious effect on behavioral variation due to a random generative process. Rats were not detrimentally affected by the retention interval; indeed, they actually demonstrated *better* performance (i.e., greater variation) with longer RIs. In a control group (e.g., Group REP) rewarded for a repetitive response sequence (e.g., LLRR), RIs interfered with performance proportional to the length of time the rat had to wait before response emission. At an RI of 20 seconds, rats in group VAR met the response criterion and were reinforced on approximately 65 percent of trials; at the same RI, rats in group REP were reinforced on only about 5 percent of trials. These results led Neuringer to hypothesize that memory for prior responses does not facilitate operant variation, but instead interferes with variation (see also Weiss, 1965). This hypothesis is consistent with earlier findings (Page & Neuringer, 1985), in which pigeons were more likely to meet a variability criterion if a required behavioral sequence was eight responses rather than four responses in length. These data are perhaps the strongest indicators for a stochastic process being responsible for operant variability. The fact that stochastic responding is controlled by its consequences is a strong indicator that it is functional (Neuringer, 2004).

Neuringer, Deiss, and Olson (2000) investigated whether the direct reinforcement of variable behavior facilitates selection of a highly rewarded target response. In Experiment 1, they trained three groups of rats in a bar-pressing task; across phases of the procedure, they arbitrarily assigned sequences of target responses across two levers, ranging from two (e.g., left-right) to five responses (e.g., right-left-left-right-left). The rats were reinforced with a food pellet anytime the target response sequence was delivered. The first group (VAR) was trained so that variable responses (i.e., relatively unlikely response sequences, of the same length as the target) were reinforced on a variable interval one-minute schedule (VI-1). The VI-1 schedule allows frequent reinforcement but includes on average a one-minute temporal gap so that reinforcement of the target sequence is detectable. A second group (ANY) was also rewarded for nontarget responses on a VI-1 schedule, but reinforcement was not contingent on variable response sequences; in short, they were permitted, but not required, to vary their sequence behavior in order to obtain food reinforcement. A third group (CON) was only reinforced for producing the target response sequence.

Although rats in all groups learned the easy (i.e., two- and three-response) sequences equally well, *only rats receiving the VAR treatment learned to produce the more difficult target behaviors* (i.e., the five-response sequence). The rats in the CON group tended to respond at very low rates during the more difficult phases, presumably because they initially were rarely reinforced for their bar-pressing behavior; rats in the ANY group continued to bar-press at high rates identical to those produced in the VAR group. Despite their high response rates, the ANY rats' acquisition of the target behavior was markedly slower than the learning demonstrated by the VAR group for four- and five-response sequences. This clearly indicates that direct reinforcement of variability facilitates acquisition of complex operant behavior (see also Grunow & Neuringer, 2002).

Arnesen (2000; as cited in Neuringer & Jensen, 2010) examined whether reinforcement of variability facilitates future problem-solving behavior. She trained a group of rats to respond variably (see Pryor et al., 1969) when they encountered various arbitrarily selected objects (e.g., a soup can). For example, a rat may have been initially reinforced for touching the soup can with its nose, but then was required to perform a different action for reinforcement (e.g., touching the top of the can with its forepaws). A separate control group was presented with the same objects, but was reinforced for interactions with the objects irrespective of behavioral variation. Rats were then tested by being individually placed in an open field with thirty objects, each of which had a small piece of food hidden on or within it. Animals that had been previously explicitly reinforced for variable behavior found and consumed significantly more hidden food than the control group. These data indicate that problem-solving behavior is made more effective when the subjects are reinforced for variable responding.

In summation, the generation of novel behavior can clearly be controlled by the direct reinforcement of variation. Animals that are reinforced for acting variably will increase the probability of engaging in highly novel acts, including what can potentially be labeled as innovative, creative behaviors (e.g., Pryor et al., 1969). Thus, although reinforcement of one or a specified set of actions produces decidedly noncreative behavior (e.g., steady-state performance of lever-pressing for food reinforcement, or any overtrained skillful act), the direct reinforcement of "novel" responding does in fact result in an increase in new actions.

Expectation and the Generation of Variable Behavior

Another way for associative conditioning to modulate behavioral variation is through generation of expectations. A multitude of evidence suggests that behavioral variation increases markedly during extinction. In other words, when a previously reinforced behavior (e.g., lever pressing) is no longer followed by reward, the action does not merely become less frequent; behavior also tends to become more variable in nature. Many researchers have taken note of this relationship (Antonitis, 1951; Balsam, Deich, Ohyama, & Stokes, 1998; Eckerman & Lanson, 1969; Herrick & Bromberger, 1965; Millenson & Hurwitz, 1961; Neuringer, Kornell, & Olufs, 2001; Notterman, 1959; Stebbins & Lanson, 1962).

There is strong evidence that this increase in the variation of behavior is dependent on the animal's Pavlovian expectation of appetitive (i.e., positive) events. Gharib, Derby, and Roberts (2001) trained rats to lever-press on a task called the peak procedure (Roberts, 1981). A trial would begin with the delivery of a discrete stimulus (e.g., a tone) that would be presented continuously throughout the trial. On 80 percent of trials, rats were reinforced with a food pellet following the first response after 40 seconds from the start of the trial. The remaining trials were probe trials, which lasted 195 seconds and terminated without food reinforcement. Gharib et al. found that the variability of rats' bar-press durations increased significantly following the point at which the animal could predict an omission of food reward on a given probe trial (i.e., after 40 seconds on a trial). These results led Gharib et al. to propose a rule: *Reduction in reward expectation increases variation in the form of behavior.* A subsequent study confirmed the rule: the interresponse times of bar-pressing behavior in rats were more variable in the presence of a discriminative stimulus that signaled a low probability of reward, as compared to a stimulus signaling a high probability of reward (Gharib, Gade, & Roberts, 2004).

Neuringer et al. (2001, Experiment 3) conducted a study to investigate operant variation and extinction. They trained rats to respond across three operanda (a key, a left lever, and a right lever). They trained two groups of rats: a VAR group was reinforced for performing response sequences that were relatively uncommon, while a REP group was required to respond in a single response sequence (e.g., left-key-right) in order to obtain reinforcement. An extinction phase was introduced following acquisition of the appropriate response strategy. Unsurprisingly, during the reinforcement phase, the VAR group responded with significantly greater variability across the operanda relative to the REP group. However, both groups were more likely to meet the variability criterion (i.e., they both responded with greater variability than in initial training) during the extinction phase. Interestingly, in extinction, there was no difference between the groups in the amount of elicited behavioral variability. An additional interesting note: the operant chambers in which the rats were trained were equipped with a pair of response keys that were never reinforced during the acquisition phase. During initial learning, rats in both groups learned that responses to these operanda were not reinforced, and quickly stopped responding. However, during the extinction phase, responding to these keys rapidly and dramatically increased. Neuringer et al. (2001) write:

> Variability increases in extinction because of the relatively large increases in low-probability behaviors. *When reinforcers are no longer forthcoming, subjects occasionally try something different.* (p. 90, emphasis added)

Our recent work has extended the examination of expectation-controlled respondent variability in behavior to open-field foraging behavior in rats (Stahlman & Blaisdell, 2011a); spatial and temporal variability of instrumental key-pecking behavior in pigeons (Stahlman, Roberts, & Blaisdell, 2010; Stahlman & Blaisdell, 2011b); and pigeons' Pavlovian conditioned key-pecking behavior (Stahlman, Young, & Blaisdell, 2010). The latter example is particularly interesting because we find that the variability of behavior is dependent on the likelihood of food delivery even when the behavior is entirely *inconsequential*. In this study, pigeons observed pairings between colored discs and grain delivery; some discs (e.g., red) were consistently followed with a relatively high probability of grain delivery (e.g., 100%), while other discs were followed with a relatively low probability of food (e.g., 1%). In pigeons, the pairing of a visual stimulus with subsequent grain delivery will typically result in an increase in pecking to the visual stimulus over cumulative training trials. Despite its impotence, pecking is acquired and maintained virtually indefinitely. In our experiment, we found that behavior was more variable in both spatial and temporal domains on trials signaling a low probability of reinforcement. This indicates that the novelty of behavior, or the distribution of behavioral outputs, is an inverse function of Pavlovian expectation of positive outcomes.

In humans, a compelling case can be made for the role of reward expectation modulating the novelty (creativity) of behavior. There is a strong association between mood disorders and human creativity. Anecdotally, some of the most creative and greatest artistic minds in recent history are those who were diagnosed with or thought to have suffered from major depressive disorder or bipolar disorder (e.g., Jackson Pollock, Sylvia Plath, Ernest Hemingway, Virginia Woolf). There is a great deal of quantified evidence for a link between mood disorders characterized by depression (i.e., major depressive disorder, manic-depressive disorder) and creative behavior (e.g., Akinola & Mendes, 2008; Andreasen, 1987; Jamison, 1989, 1997; Lauronen, Veijola, Isohanni, Jones, Nieminen, & Isohanni, 2004;

Richards, Kinney, Lunde, Benet, & Merzel, 1988; Simeonova, Chang, Strong, & Ketter, 2005). An important and pervasive characteristic of depressed individuals is that they report feeling helpless and hopeless, that nothing good will happen regardless of their actions. In an important way, this seems analogous to a situation whereby a pigeon has learned that its pecking behavior will not result in a positive outcome. As the pigeon engages in highly novel and variable behavior induced by low reinforcement expectation, so too might depressed individuals. We will cover this in more detail below, where we discuss the reward circuitry of the brain as it pertains to variation in behavior.

It makes a good deal of sense for an animal to behave with greater variability if reinforcement is unlikely. Gharib et al. (2004) describe the adaptive functionality of the relationship between variability and reward expectation well:

> If an animal's actions vary too little, it will not find better ways of doing things; if they vary too much, rewarded actions will not be repeated. So at any time there is an optimal amount of variation, which changes as the costs and benefits of variation change. Animals that learn instrumentally would profit from a mechanism that regulates variation so that the actual amount is close to the optimal amount. (p. 271)

The results of our study in a Pavlovian task with pigeons support Gharib et al.'s (2004) assertion, or at least the part of it related to the production of variable behavior. (The data pertaining to individuals with mood disorders are similarly supportive of Gharib et al.'s position.) On trials with a specifically low probability of grain delivery, the pigeons occasionally pecked on the touchscreen at a location very far from the stimulus target. Let's say that I selected a region of space (away from the touchscreen) to designate as a "secret cache" of reward, such that a peck to that off-target location produced a *certain* reward. Pigeons pecking with high variability to the screen would presumably be more likely to discover this secret cache than pigeons with a narrow distribution of response location (e.g., on high-probability trials).

Empirical evidence for the negative effect of reinforcement on behavioral variability with respect to creativity comes from studies showing that variability tends to decrease as an animal draws nearer to reward (Gharib et al., 2001; Neuringer, 1991; Schwartz, 1982). The observation that variability tends to decrease with approach to reinforcers certainly suggests that reinforcement interferes with production of novel behavior (Cherot et al., 1996). Reinforcement of variability tends to increase total levels of variability (e.g., Page & Neuringer, 1985), but as outcomes become more proximal on a given trial, variability decreases. Each of these opposing effects of reinforced variability may have implications for the production of creative behavior.

Instrumental learning requires the selection of an action from numerous alternative behavioral options. Action is selected through differential reinforcement. As the likelihood of reinforcement decreases, the selective force governing instrumental behavior tends to relax, resulting in more variety in behavior. This increase in the variety of output increases the likelihood that the animal will stumble upon a behavioral option that is more valuable in terms of its consequences, at which time the animal can reduce the variability of its response around the new action.

Neurobiology of Behavioral Variation

An examination of the neurobiology that underlies variability and stereotypy in behavior reveals a great deal of complexity in the neural architecture employed in both reward processing and in expectation generation. Nevertheless, an inspection of the literature indicates that the structures of the basal ganglia, specifically, are critical to learning and to the production of behavioral variability.

A great amount of research supports the role of the basal ganglia in variability production. Abnormalities of the cortico-basal ganglia circuits have been linked to both motor and cognitive repetition (e.g., Leckman, 2002). Furthermore, direct manipulation of brain chemistry with respect to the basal ganglia's dopaminergic structures has been found to alter the levels of behavioral variation in animals; severe stereotypies can be induced by administration of dopamine and opioid agonists (Canales & Graybiel, 2000; Saka, Goodrich, Harlan, Madras, & Graybiel, 2004; Saka, Iadarola, Fitzgerald, & Graybiel, 2002), and can be reduced by delivery of dopamine antagonists, such as haloperidol (Devenport, Devenport, & Holloway, 1981). Saka et al. (2004) found a strong correlation between activation of the striatum (especially the putamen) and cocaine-induced behavioral stereotypy in squirrel monkeys. Other experiments have found that opiate agonist-induced disinhibition of nigrostriatal dopaminergic projections induces strong stereotypy in rats (Iwamoto & Way, 1977). Ultimately, drug-induced stereotypies seem to result from behavioral disinhibition caused by abnormal functioning of the dorsal striatum and, by extension, an imbalance of neural activation in favor of the striatonigral pathway of the basal ganglia. Specifically, pharmacological manipulations that increase the activity of the direct, striatonigral pathway of the basal ganglia increase motor stereotypy; similarly, lesions or pharmacological down-regulations of the activity of the indirect, striatopallidal pathway of the basal ganglia induce stereotypy (Garner & Mason, 2002; Lewis & Kim, 2009; Presti & Lewis, 2005). Conversely, the administration of dopaminergic agonists that selectively target the striatopallidal pathway increases variability in behavior (Longoni et al., 1991). J. P. Garner comments on the competition between basal ganglia pathways with respect to variable and stereotyped behavior: "Broadly speaking, stereotypy can thus be reduced by drugs that activate the indirect pathway or suppress the direct pathway; while stereotypy is selectively induced by drugs that suppress the indirect pathway" (in Lewis, Presti, Lewis, & Turner, 2006, p. 210).

Stereotypic behavior can be elicited in other ways besides through the administration of psychoactive stimulants; for example, captive animals frequently exhibit spontaneous stereotypic behavior. These "cage stereotypies" are thought to be the product of stress, coupled with low behavioral competition due to environmental conditions lacking in complexity. Recent work in deer mice indicates that animals with relatively low activity in the subthalamic nucleus (a component of the indirect pathway of the basal ganglia) show higher rates of cage stereotypies (Tanimura, King, Williams, & Lewis, 2011); environmental enrichment both reduces the amount of stereotypic behavior and normalizes the activity of the subthalamic nucleus. Studies indicate that dysfunction and abnormalities of the basal ganglia are also related to spontaneous stereotypic behavior in laboratory rats (Garner & Mason, 2002), horses (McBride & Hemmings, 2005), parrots (Garner, Meehan, & Mench, 2003), and humans (e.g., Mink & Pleasure, 2003). The neurobiological correlates of stereotypic behavior seem to be conserved across phylogenetically distant species.

There has been a recent recognition of the role of the reward circuitry of the brain and its relationship to depression in humans. For example, Nestler and Carlezon (2006) present data that the behavioral symptomology of depression is, at least in part, due to malfunctioning of the striatum and ventral tegmentum, both of which are critical components of the basal ganglia circuit. There is mounting evidence that the structures of the basal ganglia are implicated in behavior relevant to depressive disorder in humans (Krishnan et al., 2007; Krishnan & Nestler, 2008). As we discussed above, an individual who displays depressive symptomology (as in major depressive disorder or bipolar disorder) is more likely than the general population to be involved in creative work (e.g., as an artist). It is therefore not surprising to find that the neural regions that are implicated in the production of stereotypic and variable behavior are also implicated in depression.

Study of spontaneous and drug-induced stereotypic behavior is important for our understanding of the role of the basal ganglia in producing stereotypic behavior. However, our primary focus in this chapter has been on behavior in associative preparations; we are interested in variation produced by direct reinforcement of behavioral variation (e.g., Page & Neuringer, 1985) and variation driven by reductions in the predicted amount of reinforcement (e.g., Stahlman et al., 2010a). Recently, a large amount of research has been devoted to investigating the role of the basal ganglia in instrumental behavior; normal functioning of the components of the ganglia appear to be critical for normal functioning in instrumental learning procedures. The nucleus accumbens (e.g., Hernandez, Sadeghian, & Kelley, 2002; Koch, Schmid, & Schnitzler, 2000; Salamone, Correa, Farrar, & Mingote, 2007; Wyvell & Berridge, 2000) and the striatum (e.g., Wiltgen, Law, Ostlund, Mayford, & Balleine, 2007; Yin & Knowlton, 2006; Yin, Knowlton, & Balleine, 2006) seem to be important for the acquisition and performance of instrumental actions.

A rapidly increasing body of literature suggests that the basal ganglia are of principal importance both for expectation and in regulating behavioral variability particularly in associative preparations. In rhesus macaques, information regarding the size of an expected reward is encoded by neurons in the anterior striatum (Cromwell & Schultz, 2003); other studies have demonstrated that motor behavior in monkeys is shaped by incentive value encoded in the basal ganglia circuit (Pasquereau et al., 2007), and that neural activity in the caudate nucleus accurately predicts both rewarded and unrewarded action (Watanabe, Lauwereyns, & Hikosaka, 2003). There is evidence to support the role for prefrontal cortical structures in modulating the behavior in the striatum during reward encoding (Staudinger, Erk, & Walter, 2011). Graybiel (2005), among others, has suggested that reinforcement signals (e.g., magnitude and likelihood of reward) are instantiated in the basal ganglia. In addition, she suggests that the basal ganglia are critically important in maintaining the balance of exploration and exploitation in conditioned animal behavior, thereby optimizing response output to the expected conditions of reward. It is important to note the confluence of this suggestion with Gharib et al.'s (2004) argument that variability in behavior must be appropriately modulated as the costs and benefits of variation change.

Neurobiological evidence from songbirds indicates that the functionality of the basal ganglia with respect to production of variation is conserved across even phylogenetically distant relatives. Brainard & Doupe (2000) discovered that lesions of the lateral magnocellular nucleus of the anterior nidopallium (LMAN, an avian cortical-basal ganglia circuit) result in unusual stereotypy in song in male zebra finches. Recent studies (Brainard & Doupe, 2001; Kao & Brainard, 2006; Kao, Doupe, & Brainard, 2005) have corroborated the existence of a positive correlation between variation in song and activity of the LMAN in male zebra finches. As a zebra finch male ages, the activity of the LMAN, the variation in its song output, and its ability to modulate its song all decrease (Kao & Brainard, 2006). These findings support the hypothesis that the basal ganglia are critical for the production and modulation of song variation; importantly, this variability in song is adaptive (i.e., not mere noise), allowing finches to rapidly learn to shift the pitch of their songs to avoid an external disruptor (Tumer & Brainard, 2007). Dopaminergic connections within the circuitry of the basal ganglia are critically important to modulate song variability in adult songbirds (Leblois, Wendel, & Perkel, 2010). The adaptability of the control of variability in behavior is not confined to songbirds; indeed, research with mice (Tanimura, Yang, & Lewis, 2008), voles (Garner & Mason, 2002), bears (Vickery & Mason, 2005), and humans (e.g., Neuringer, 2002) confirms the relationship of variability production with the generation of adaptive action.

Conclusions

The topics covered in this chapter would seem to be important pieces of the creativity puzzle. The individual must be able to engage in novel action in order for creativity to be a possibility. As behavior must be in some way novel to be described as creative, the animal must possess mechanisms related to the production of novel action. In this chapter, we have described associative processes as being heavily involved in the induction of novel actions. Theoretical accounts of expectation-induced variability (and stereotypy) in behavior suggest that animals engage in novel or unusual behavior in situations where they have learned that positive reinforcement is unlikely (Gharib et al., 2001). Similarly, direct reinforcement of variation in behavior will produce an animal that has a wider range of action, approaching a random distribution of the response measure (Neuringer, 2004). Indeed, when the variability of its behavior *itself* is reinforced, the animal may engage in such novel, random, and/or complex actions that they become functionally impossible to keep track of (Pryor et al., 1969). Importantly, operant variability appears to reflect instrumental control of a stochastic process, rather than the dynamics of a memory process.

In summary, we have much to learn about the production of novel and variable action by examining nonhuman animals. Associative conditioning preparations can engender high variability in behavior, which provides the animal with additional options for action. The engine that produces novelty in behavior is related to the interplay of the animal's ability to predict (1) whether reinforcement will occur, and 2) what manner of response will be followed by reinforcement. Any complete account of creative behavior should incorporate these generative processes as a foundation upon which other processes can be built.

Acknowledgments

Support for the production of this chapter was provided by NIH Grant NS059076 (A. P. Blaisdell). Requests for reprints should be addressed to Aaron P. Blaisdell, UCLA Department of Psychology, 1285 Franz Hall, Box 951563, Los Angeles, CA 90095–1563, USA; e-mail: blaisdell@psych.ucla.edu.

Note

1. There are many other facets relevant to a discussion of creativity in animals (including the human animal) beyond the mechanisms discussed in this chapter: the directed combination of separately learned behaviors or "insight" (e.g., Epstein, Kirshnit, Lanza, & Rubin, 1984; Kohler, 1925); inferences and the spontaneous integration of spatiotemporal maps (Blaisdell, 2009; Leising & Blaisdell, 2009; Leising, Sawa, & Blaisdell, 2007; Sawa, Leising, & Blaisdell, 2005); and causal reasoning (Blaisdell, Sawa, Leising, & Waldmann, 2006), to just name a few.

References

Akinola, M., & Mendes, W. B. (2008). The dark side of creativity: Biological vulnerability and negative emotions lead to greater artistic creativity. *Personality and Social Psychology Bulletin, 34,* 1677–1686.

Andreasen, N. C. (1987). Creativity and mental illness: Prevalence rates in writers and their first-degree relatives. *American Journal of Psychiatry, 144,* 1288–1292.

Antonitis, J. J. (1951). Response variability in the white rat during conditioning, extinction, and reconditioning. *Journal of Experimental Psychology, 42,* 273–281.

Arnesen, E. M. (2000). *Reinforcement of object manipulation increases discovery.* Unpublished undergraduate thesis, Reed College.

Balsam, P. D., Deich, J. D., Ohyama, T., & Stokes, P. D. (1998). *Origins of new behavior.* Needham Heights, MA: Allyn & Bacon.

Bartlett, R., Wheat, J., & Robins, M. (2007). Is movement variability important for sports biomechanists? *Sports Biomechanics, 6*, 224–243.

Blaisdell, A. P. (2009). The role of associative processes in spatial, temporal, and causal cognition. In S. Watanabe, A. P. Blaisdell, L. Huber, & A. Young (Eds.), *Rational animals, irrational humans* (pp. 153–172). Tokyo: Keio University.

Blaisdell, A. P., Sawa, K., Leising, K. J., & Waldmann, M. R. (2006). Causal reasoning in rats. *Science, 311*, 1020–1022.

Blough, D. S. (1966). The reinforcement of least-frequent interresponse times. *Journal of the Experimental Analysis of Behavior, 9*, 581–591.

Boden, M. A. (2004). *The creative mind: Myths and mechanisms.* New York: Routledge.

Brainard, M. S., & Doupe, A. J. (2000). Interruption of a basal ganglia-forebrain circuit prevents plasticity of learned vocalizations. *Nature, 404*, 762–766.

Brainard, M. S., & Doupe, A. J. (2001). Postlearning consolidation of birdsong: Stabilizing effects of age and anterior forebrain lesions. *Journal of Neuroscience, 21*, 2501–2517.

Brembs, B. (2011). Towards a scientific concept of free will as a biological trait: Spontaneous actions and decision-making in invertebrates. *Proceedings of the Royal Society of London, Series B: Biological Sciences, 278*, 930–939.

Bryant, D., & Church, R. M. (1974). The determinants of random choice. *Animal Learning & Behavior, 2*, 245–248.

Canales, J. J., & Graybiel, A. M. (2000). A measure of striatal function predicts motor stereotypy. *Nature Neuroscience, 3*, 377–383.

Cherot, C., Jones, A., & Neuringer, A. (1996). Reinforced variability decreases with approach to reinforcers. *Journal of Experimental Psychology: Animal Behavior Processes, 22*, 497–508.

Cohen, L., Neuringer, A., & Rhodes, D. (1990). Effects of ethanol on reinforced variations and repetitions by rats under a multiple schedule. *Journal of the Experimental Analysis of Behavior, 54*, 1–12.

Cook, R. G., Brown, M. F., & Riley, D. A. (1985). Flexible memory processing by rats: Use of prospective and retrospective information in the radial maze. *Journal of Experimental Psychology: Animal Behavior Processes, 11*, 453–469.

Cook, R. G., Levison, D. G., Gillett, S. R., & Blaisdell, A. P. (2005). Capacity and limits of associative memory in pigeons. *Psychonomic Bulletin & Review, 12*, 350–358.

Cromwell, H. C., & Schultz, W. (2003). Effects of expectations for different reward magnitudes on neuronal activity in primate striatum. *Journal of Neurophysiology, 89*, 2823–2838.

Dennett, D. C. (1975). Why the law of effect will not go away. *Journal for the Theory of Social Behaviour, 5,* 169–188.

Denney, J., & Neuringer, A. (1998). Behavioral variability is controlled by discriminative stimuli. *Animal Learning & Behavior, 26,* 154–162.

Devenport, L. D., Devenport, J. A., & Holloway, F. A. (1981). Reward-induced stereotypy: Modulation by the hippocampus. *Science, 212,* 1288–1289.

Eckerman, D. A., & Lanson, R. N. (1969). Variability of response location for pigeons responding under continuous reinforcement, intermittent reinforcement, and extinction. *Journal of the Experimental Analysis of Behavior, 12,* 73–80.

Epstein, R. (1990). Generativity theory and creativity. In M. A. Runco & R. S. Albert (Eds.), *Theories of Creativity* (pp. 116–140). Thousand Oaks, CA: Sage Publications.

Epstein, R. (1991). Skinner, creativity, and the problem of spontaneous behavior. *Psychological Science, 2,* 362–370.

Epstein, R., Kirshnit, C. E., Lanza, R. P., & Rubin, L. C. (1984). "Insight" in the pigeon: Antecedents and determinants of an intelligent performance. *Nature, 308,* 61–62.

Garner, J. P., Meehan, C. L., & Mench, J. A. (2003). Stereotypies in caged parrots, schizophrenia, and autism: Evidence for a common mechanism. *Behavioural Brain Research, 145,* 125–134.

Garner, J. P., & Mason, G. J. (2002). Evidence for a relationship between cage stereotypies and behavioural disinhibition in laboratory rodents. *Behavioural Brain Research, 136,* 83–92.

Gharib, A., Gade, C., & Roberts, S. (2004). Control of variation by reward probability. *Journal of Experimental Psychology: Animal Behavior Processes, 30,* 271–282.

Gharib, A., Derby, S., & Roberts, S. (2001). Timing and the control of variation. *Journal of Experimental Psychology: Animal Behavior Processes, 27,* 165–178.

Graybiel, A. M. (2005). The basal ganglia: Learning new tricks and loving it. *Current Opinion in Neurobiology, 15,* 638–644.

Grunow, A., & Neuringer, A. (2002). Learning to vary and varying to learn. *Psychonomic Bulletin & Review, 9,* 250–258.

Hernandez, P. J., Sadeghian, K., & Kelley, A. E. (2002). Early consolidation of instrumental learning requires protein synthesis in the nucleus accumbens. *Nature Neuroscience, 5,* 1327–1331.

Herrick, R. M., & Bromberger, R. A. (1965). Lever displacement under a variable ratio schedule and subsequent extinction. *Journal of Comparative and Physiological Psychology, 59,* 392–398.

Iwamoto, E. T., & Way, E. L. (1977). Circling behavior and stereotypy induced by intranigral opiate microinjections. *Journal of Pharmacology and Experimental Therapeutics, 203,* 347–359.

Jamison, K. R. (1989). Mood disorders and patterns of creativity in British writers and artists. *Psychiatry: Journal for the Study of Interpersonal Processes, 52,* 125–134.

Jamison, K. R. (1997). Manic-depressive illness and creativity. *Scientific American* (February): 44–49.

Kao, M. H., & Brainard, M. S. (2006). Lesions of an avian basal ganglia circuit prevent context-dependent changes to song variability. *Journal of Neurophysiology, 96,* 1441–1455.

Kao, M. H., Doupe, A. J., & Brainard, M. S. (2005). Contributions of an avian basal ganglia-forebrain circuit to real-time modulation of song. *Nature, 433,* 638–643.

Kesner, R. P., & Despain, M. J. (1988). Correspondence between rats and humans in the utilization of retrospective and prospective codes. *Animal Learning & Behavior, 16,* 299–302.

Koch, M., Schmid, A., & Schnitzler, H. U. (2000). Role of nucleus accumbens dopamine D1 and D2 receptors in instrumental and Pavlovian paradigms of conditioned reward. *Psychopharmacology, 152,* 67–73.

Kohler, W. (1925). *The mentality of apes*. London: Routledge & Kegan Paul.

Krishnan, V., Han, M. H., Graham, D. L., Berton, O., Renthal, W., Russo, S. J., et al. (2007). Molecular adaptations underlying susceptibility and resistance to social defeat in brain reward regions. *Cell, 131,* 391–404.

Krishnan, V., & Nestler, E. J. (2008). The molecular neurobiology of depression. *Nature, 455,* 894–902.

Lauronen, E., Veijola, J., Isohanni, I., Jones, P. B., Nieminen, P., & Isohanni, M. (2004). Links between creativity and mental disorder. *Psychiatry, 67,* 81–98.

Leblois, A., Wendel, B. J., & Perkel, D. J. (2010). Striatal dopamine modulates basal ganglia output and regulates social context-dependent behavioral variability through D_1 receptors. *Journal of Neuroscience, 30,* 5730–5743.

Leckman, J. (2002). Tourette's syndrome. *Lancet, 360,* 1577–1586.

Leising, K. J., & Blaisdell, A. P. (2009). Associative basis of landmark learning and integration in vertebrates. *Comparative Cognition & Behavior Reviews, 4,* 80–102.

Leising, K. J., Sawa, K., & Blaisdell, A. P. (2007). Temporal integration in Pavlovian appetitive conditioning in rats. *Learning & Behavior, 35,* 11–18.

Lewis, M., & Kim, S. (2009). The pathophysiology of restricted repetitive behavior. *Journal of Neurodevelopmental Disorders, 1,* 114–132.

Lewis, M. H., Presti, M. F., Lewis, J. B., & Turner, C. A. (2006). The neurobiology of stereotypy. In G. Mason & J. Rushen (Eds.), *Stereotypic animal behaviour: Fundamentals and applications to welfare* (2nd Ed., pp. 190–226). Cambridge, MA: CABI.

Longoni, R., Spina, L., Mulas, A., Carboni, E., Garau, L., Melchiorri, P., et al. (1991). (d-Ala2)deltorphin II: D1-dependent stereotypies and stimulation of dopamine release in the nucleus accumbens. *Journal of Neuroscience, 11,* 1565–1576.

McBride, S. D., & Hemmings, A. (2005). Altered mesoaccumbens and nigro-striatal dopamine physiology is associated with stereotypy development in a non-rodent species. *Behavioural Brain Research, 159,* 113–118.

McElroy, E., & Neuringer, A. (1990). Effects of alcohol on reinforced repetitions and reinforced variations in rats. *Psychopharmacology, 102,* 49–55.

Millenson, J. R., & Hurwitz, H. M. B. (1961). Some temporal and sequential properties of behavior during conditioning and extinction. *Journal of the Experimental Analysis of Behavior, 4,* 97–106.

Mink, J. W., & Pleasure, D. E. (2003). The basal ganglia and involuntary movements: Impaired inhibition of competing motor patterns. *Archives of Neurology, 60,* 1365–1368.

Morgan, L., & Neuringer, A. (1990). Behavioral variability as a function of response topography and reinforcement contingency. *Animal Learning & Behavior, 18,* 257–263.

Nestler, E. J., & Carlezon, W. A., Jr. (2006). The mesolimbic dopamine reward circuit in depression. *Biological Psychiatry, 59,* 1151–1159.

Neuringer, A. (1986). Can people behave "randomly"? The role of feedback. *Journal of Experimental Psychology: General, 115,* 62–75.

Neuringer, A. (1991). Operant variability and repetition as functions of interresponse time. *Journal of Experimental Psychology: Animal Behavior Processes, 17,* 3–12.

Neuringer, A. (1993). Reinforced variation and selection. *Animal Learning & Behavior, 21,* 83–91.

Neuringer, A. (2002). Operant variability: Evidence, functions, and theory. *Psychonomic Bulletin & Review, 9,* 672–705.

Neuringer, A. (2004). Reinforced variability in animals and people: Implications for adaptive action. *American Psychologist, 59,* 891–906.

Neuringer, A., Deiss, C., & Olson, G. (2000). Reinforced variability and operant learning. *Journal of Experimental Psychology: Animal Behavior Processes, 26,* 98–111.

Neuringer, A., & Jensen, G. (2010). Operant variability and voluntary action. *Psychological Review, 117,* 972–993.

Neuringer, A., Kornell, N., & Olufs, M. (2001). Stability and variability in extinction. *Journal of Experimental Psychology: Animal Behavior Processes, 27,* 79–94.

Notterman, J. M. (1959). Force emission during bar pressing. *Journal of Experimental Psychology, 58,* 341–347.

Page, S., & Neuringer, A. (1985). Variability is an operant. *Journal of Experimental Psychology: Animal Behavior Processes, 11,* 429–452.

Pasquereau, B., Nadjar, A., Arkadir, D., Bezard, E., Goillandeau, M., Bioulac, B., et al. (2007). Shaping of motor responses by incentive values through the basal ganglia. *Journal of Neuroscience, 27,* 1176–1183.

Presti, M. F., & Lewis, M. H. (2005). Striatal opioid peptide content in an animal model of spontaneous stereotypic behavior. *Behavioural Brain Research, 157,* 363–368.

Pryor, K. W., Haag, R., & O'Reilly, J. (1969). The creative porpoise: Training for novel behavior. *Journal of the Experimental Analysis of Behavior, 12,* 653–661.

Richards, R., Kinney, D. K., Lunde, I., Benet, M., & Merzel, A. P. C. (1988). Creativity in manic-depressives, cyclothymes, their normal relatives, and control subjects. *Journal of Abnormal Psychology, 97,* 281–288.

Roberts, S. (1981). Isolation of an internal clock. *Journal of Experimental Psychology: Animal Behavior Processes, 7,* 242–268.

Ross, C., & Neuringer, A. (2002). Reinforcement of variations and repetitions along three independent response dimensions. In *Behavioural Processes, Special Issue: Proceedings of the Meeting of the Society for the Quantitative Analyses of Behavior (SQAB 2001), 57*(2–3), 199–209.

Salamone, J. D., Correa, M., Farrar, A., & Mingote, S. M. (2007). Effort-related functions of nucleus accumbens dopamine and associated forebrain circuits. *Psychopharmacology, 191,* 461–482.

Saka, E., Goodrich, C., Harlan, P., Madras, B. K., & Graybiel, A. M. (2004). Repetitive behaviors in monkeys are linked to specific striatal activation patterns. *Journal of Neuroscience, 24,* 7557–7565.

Saka, E., Iadarola, M., Fitzgerald, D. J., & Graybiel, A. M. (2002). Local circuit neurons in the striatum regulate neural and behavioral responses to dopaminergic stimulation. *Proceedings of the National Academy of Sciences, 99,* 9004–9009.

Sawa, K., Leising, K. J., & Blaisdell, A. P. (2005). Sensory preconditioning in spatial learning using a touch screen task in pigeons. *Journal of Experimental Psychology: Animal Behavior Processes, 31,* 368–375.

Schoenfeld, W. N., Harris, A. H., & Farmer, J. (1966). Conditioning response variability. *Psychological Reports, 19,* 551–557.

Schwartz, B. (1982). Interval and ratio reinforcement of a complex sequential operant in pigeons. *Journal of the Experimental Analysis of Behavior, 37,* 349–357.

Simeonova, D. I., Chang, K. D., Strong, C., & Ketter, T. A. (2005). Creativity in familial bipolar disorder. *Journal of Psychiatric Research, 39,* 623–631.

Skinner, B. F. (1966). An operant analysis of problem solving. In B. Kleinmuntz (Ed.), *Problem solving: Research, method, and theory* (pp. 225–257). New York: John Wiley.

Skinner, B. F. (1970). Creating the creative artist. In A. J. Toynbee, et al. (Eds.), *On the future of art* (pp. 61–75). New York: Viking Press.

Stahlman, W. D., & Blaisdell, A. P. (2011a). Reward probability and the variability of foraging behavior in rats. *International Journal of Comparative Psychology, 24,* 168–176.

Stahlman, W. D., & Blaisdell, A. P. (2011b). The modulation of response variation by the probability, magnitude, and delay of reinforcement. *Learning and Motivation, 42,* 221–236.

Stahlman, W. D., Roberts, S., & Blaisdell, A. P. (2010a). Effect of reward probability on spatial and temporal variation. *Journal of Experimental Psychology: Animal Behavior Processes, 36,* 77–91.

Stahlman, W. D., Young, M. E., & Blaisdell, A. P. (2010b). Response variability in pigeons in a Pavlovian task. *Learning & Behavior, 38,* 111–118.

Staudinger, M. R., Erk, S., & Walter, H. (2011). Dorsolateral prefrontal cortex modulates striatal reward encoding during reappraisal of reward anticipation. *Cerebral Cortex, 21,* 2578–2588.

Stebbins, W. C., & Lanson, R. N. (1962). Response latency as a function of reinforcement schedule. *Journal of the Experimental Analysis of Behavior, 5,* 299–304.

Tanimura, Y., King, M. A., Williams, D. K., & Lewis, M. H. (2011). Development of repetitive behavior in a mouse model: Roles of indirect and striosomal basal ganglia pathways. *International Journal of Developmental Neuroscience, 29,* 461–467.

Tanimura, Y., Yang, M. C., & Lewis, M. H. (2008). Procedural learning and cognitive flexibility in a mouse model of restricted, repetitive behavior. *Behavioural Brain Research, 189,* 250–256.

Thorndike, E. L. (1927). The law of effect. *American Journal of Psychology, 39,* 212–222.

Tumer, E. C., & Brainard, M. S. (2007). Performance variability enables adaptive plasticity of crystallized adult birdsong. *Nature, 450,* 1240–1244.

Vickery, S. S., & Mason, G. J. (2005). Stereotypy and perseverative responding in caged bears: Further data and analyses. *Applied Animal Behaviour Science, 91,* 247–260.

Watanabe, K., Lauwereyns, J., & Hikosaka, O. (2003). Neural correlates of rewarded and unrewarded eye movements in the primate caudate nucleus. *Journal of Neuroscience, 23,* 10052–10057.

Weiss, R. L. (1965). Variables that influence random-generation: An alternative hypothesis. *Perceptual and Motor Skills, 20,* 307–310.

Wiltgen, B. J., Law, M., Ostlund, S., Mayford, M., & Balleine, B. W. (2007). The influence of Pavlovian cues on instrumental performance is mediated by CaMKII activity in the striatum. *European Journal of Neuroscience, 25,* 2491–2497.

Wyvell, C. L., & Berridge, K. C. (2000). Intra-accumbens amphetamine increases the conditioned incentive salience of sucrose reward: Enhancement of reward "wanting" without enhanced "liking" or response reinforcement. *Journal of Neuroscience, 20,* 8122–8130.

Yin, H. H., & Knowlton, B. J. (2006). The role of the basal ganglia in habit formation. *Nature Reviews: Neuroscience, 7,* 464–476.

Yin, H. H., Knowlton, B. J., & Balleine, B. W. (2006). Inactivation of dorsolateral striatum enhances sensitivity to changes in the action-outcome contingency in instrumental conditioning. *Behavioural Brain Research, 166,* 189–196.

II Genetics

4 The Genetics of Creativity: The Generative and Receptive Sides of the Creativity Equation

Baptiste Barbot, Mei Tan, and Elena L. Grigorenko

Art is the social within us, and even if its action is performed by a single individual, it does not mean that its essence is individual.
—Lev Vygotsky (1971, p. 249)

Creative processes, products, or persons have been conceptualized in many ways across many domains, and many attempts have been made to characterize these uniquely human phenomena from diverse perspectives—religious and mystic, philosophical, sociological, psychological, and historical. Hence, there are hundreds of definitions of creativity (e. g., Treffinger, 1996), but an increasingly consensual definition is that creativity is the ability to produce something that is both new/original and adapted/appropriate/valuable to a particular task or domain (e.g., Amabile, 1996; Barron, 1988; MacKinnon, 1962; Ochse, 1990; Sternberg & Lubart, 1995). Thus, as an "ability," creativity is a phenomenon originating in individuals. However, because creativity results in a product that is valued to different degrees by an audience—appreciated, rejected, embraced, and cultivated, or put aside and forgotten—creativity is also a sociocultural phenomenon. While many studies have looked into the psychological factors contributing to the individual's creativity, including aspects of both cognition (e.g., divergent thinking and intelligence) and conation (e.g., personality and motivational factors) that lead to the development of creative products (Selby, Shaw, & Houtz, 2005), the making of creative products is only part of the "creativity equation," and few authors have looked at the other part—the cultural and social environment in which and for which the creative product emerges (e.g., Lubart, 1990).

The influence of genetic factors on the first part of the creativity equation is increasingly being investigated, with several studies focused on the genetic and neurobiological sources of this individual side of creativity (Reuter, Roth, Holve, & Hennig, 2006). However, the genetic factors that may influence the environmental side of creativity—that is, the genetic bases that underlie the reception of creative products—as well as the processes by which creative products may impact the human genome, have rarely been mapped out. This chapter does not pretend to do so, but aims to point out and discuss some of the important issues of the research into the genetic contributions to creativity that are generally disregarded and need further exploration, specifically: What are the societal/environmental factors that can affect a culture's response to novel products or ideas? What are the possible genetic roots of these social processes that, to some extent, define what is creative or not and determine the "adoption" of new products? The answers to these questions will help us more fully understand the nature of creativity and its function in culture and society, leading to better ways to encourage creativity, both by nurturing individuals as creators and by providing cultural environments conducive to creative behavior.

To address these questions, we first provide some background on the current state of knowledge on the genetic etiology of creativity, which thus far refers essentially to the genetic bases of the individual factors involved in creative ability. Second, we review the sociocultural aspects of creativity and present how these aspects have been interpreted from an evolutionary perspective. This includes examinations of the genetic influences that shape populations and their cultures, the cultural environment that receives the creative product and determines its usefulness and value, and some possible interactive effects that may contribute both to creativity as an individual ability and to the reception and adoption of novelty as a social process. We conclude by underlining the importance of studying creativity not only as an individual, objective "ability," but also as a cultural, time-specific, biologically grounded phenomenon with a social purpose.

Facets of Creativity

Creativity as Ability: The Genetic Basis of the "Creative Individual"

The idea that "genius runs in families," suggesting the hereditary nature of creative giftedness and talent, can be traced back at least to Francis Galton (Guilford, 1987). Nowadays, there is substantial interest in understanding the genetic bases of creativity and related constructs (e.g., Chavez-Eakle, Graff-Guerrero, García-Reyna, Vaugier, & Cruz-Fuentes, 2007; Kaufman, Kornilov, Bristol, Tan, & Grigorenko, 2010), interpreted as individual ability. This approach and a related growing body of literature on this topic (e.g., Csikszentmihalyi, 1997; Guilford, 1950; Runco, 2004; Sternberg, 2006) contribute to the understanding of both the individual and the social processes involved, which we will describe further.

Several lines of evidence support the hypothesis connecting the variation in people's genomes to the variation in people's creativity. Research focusing on the genetic bases of creativity within the traditional behavior-genetic framework—quantitative-genetic research using twin (e.g., Grigorenko, LaBuda, & Carter, 1992; Reznikoff, Domino, Bridges, & Honeyman, 1973) and family studies (e.g., Dacey, 1989; Scheinfeld, 1973; Vernon, 1989)—generally produces contradictory findings (Kaufman et al., 2010), and it is thus difficult to draw consistent conclusions from this approach. In general, these studies have produced low to moderate heritability estimates of creativity. These mixed findings could be related to the different indicators of creativity used in different studies, since different aspects of creativity (e.g., divergent-exploratory thinking processes as opposed to convergent-integrative thinking processes; see Barbot, Besançon, & Lubart, 2011) may be influenced by different genetic bases. This is consistent with the idea that creativity is an emergent property of the synergistic interaction among a cluster of more fundamental characteristics, rather than a single trait in itself (Estes & Ward, 2002). As a consequence, the genetic sources of individual differences in creativity may be partly explained through work on the genetic bases of cognition (intelligence) and conation (personality), as both contribute to individuals' ability to be creative.

The Genetic Bases of Intelligence and Other Cognitive Factors Contributing to Creativity

There is a substantial body of literature on the genetic bases of intelligence, indicating that intelligence is highly heritable (Davies et al., 2011). Indeed, 50 percent of the variance in intelligence may be explained by genetic contributions (Deary, Spinath, & Bates, 2006; Devlin, Daniels, & Roeder, 1997; Patrick, 2000; Plomin & Spinath, 2004). This has been illustrated by multiple twin and adoption studies suggesting that additive genetic effects contribute to over half of the population variance in intelligence in adults (Davies, et al., 2011; Deary, Johnson, & Houlihan, 2009; Deary, Penke, & Johnson, 2010). Across six genome-wide association scans for genes contributing to intelligence and cognition (Butcher et al., 2005; Buyske et al., 2006; Dick et al., 2006; Luciano et al., 2006; Posthuma et al., 2005; Wainwright et al., 2006), signals did cluster in regions on three chromosomes (2q, 6p, and 14q); however, the use of different measures to determine phenotype, overlapping datasets between the studies approached with different methodologies, and small effect sizes suggest that we should be cautious concerning the stability and replicability of these results (Mandelman & Grigorenko, 2011). Hence, no specific gene or gene variants have been robustly associated with phenotypes of intelligence. The most recent genome-wide association study (GWAS) for cognitive ability, conducted with five cohorts of relatively healthy middle-aged to older adults, concluded that a substantial proportion of the variance in intelligence is associated with common single nucleotide polymorphisms (SNPs, the most common type of polymorphism in the human genome, which represents the substitution of ancestral nucleotides with alternative nucleotides) in linkage disequilibrium with causal variants. This is consistent with a highly polygenic model that suggests that many genes of small effects underlie the additive genetic influence on intelligence (Davies et al., 2011).

Researchers have also recently been exploring the heritability of various creativity-related cognitive components. One such component related to creativity is divergent thinking. Kéri (2009) identified a polymorphism (rs6994992) in the promoter region of the neuregulin 1 gene to be associated with creativity in individuals with high intellectual and academic performance. The highest creative achievements and divergent-thinking scores were found in people who carried the TT genotype at rs6994992 of the neuregulin 1 gene, previously shown to be related to risk for psychosis and altered patterns of prefrontal activation. These results supplement the evidence for the possible underlying genetic basis of creativity and certain types of neuropsychiatric conditions (e.g., Smalley, Loo, Yang, & Cantor, 2005), which we will review further. Complementarily, Volf and colleagues (2009) found a significant association between verbal and figural divergent-thinking scores and the 5-HTTLPR polymorphism of the neurotransmitter serotonin transporter gene (5-HTT). The subjects with SS (i.e., homozygous for the short allele) and LS (i.e., heterozygous) genotypes demonstrated higher verbal creativity scores than the LL (i.e., homozygous for the long allele) genotype carriers. The carriers of the SS genotype also demonstrated higher figural creativity scores than the carriers of the LS and LL genotypes. According to Volf and colleagues (2009), this study provides evidence of the involvement of the central serotonin system in creativity regulation.

The Genetic Bases of Conative Factors Involved in Creativity

Researchers have also carried out many studies on the heritability of personality traits and other conative factors related to creativity. Among the key personality traits that have been examined with respect to creativity, such as risk taking (e.g., Roe et al., 2009) and openness to experience (McCrae, 1987), sensation seeking (sometimes referred to as novelty seeking) is a factor often associated with creative behavior and responses to creative products. According to Zuckerman (1994), sensation-seeking is a primary drive in both animals and humans. Neurocognitive and neuropsychological insights that could lead to a better understanding of the processes of novelty seeking and novelty finding have been reviewed (Schweizer, 2006) and confirm Zuckerman's assertion. It has indeed been argued that novelty seeking behavior is modulated by the action of the neurotransmitter dopamine (Cloninger, 1994). Specific genes involved in the substantiation of this transmission (in particular the dopamine receptor genes 4, *DRD4*, and 2, *DRD2*, and the dopamine transporter gene, *SLC6A3*) have been identified through genetic association studies and further been labeled "novelty-seeking genes" (Benjamin et al., 1996; Ebstein, Nemanov, Klotz, Gritsenko, & Belmaker, 1997; Ebstein et al., 1996; Lerman et al., 1999; Prolo & Licinio, 2002). Genetic linkage studies for novelty seeking have been carried out (Curtis, 2004), but the results are somewhat inconsistent. Munafo et al.'s (2008) meta-analysis concluded that the *DRD4* gene (specifically, its C521T polymorphism) may explain up to 3 percent of the phenotypic variance in traits of novelty seeking and impulsivity, but that the findings may be distorted by publication bias. More recently, Verweij and colleagues (2010) conducted a GWAS on harm avoidance, novelty seeking, reward dependence, and persistence using Cloninger's temperament scales (Cloninger, Przybeck, & Svrakic, 1991). The scores of a sample of 5,117 individuals were tested for association with 1,252,387 genetic markers; however, no genome-wide significant SNPs for any of the four scales were detected. This result suggests that the genetic contributions to personality consist of either many common variants of very small effect size or rare variants of somewhat larger effects, or both.

Creativity and Madness: Another Line of Evidence for the Heritable Nature of Creativity

An important body of studies has examined the heritable nature of the association between creativity and madness, proposing that a genetic component underlies this association (e.g., Glazer, 2009; see also Carson, current volume). For instance, Karlsson (1970) observed that the offspring of psychiatric patients are twice as likely as normal individuals to work in creative professional fields. Along with similar studies, these results have further cemented the conviction within the field of a link between mental illness and creativity (Glazer, 2009), with some directional hypotheses explaining this link (e.g., Richards & Kinney, 2000). Similarly, it has been hypothesized that some personality traits that are thought to be predispositional or characteristic of creativity such as psychoticism (e.g., Eysenck, 1983) might be at the same time vulnerability factors for psychopathology.

By extension, researchers have proposed that there could be some shared genetic mechanisms that contribute to such shared manifestations of creativity and mental illness (e.g., Folley, Doop, & Park, 2003), and this hypothesis has also been supported by findings from molecular-genetic studies of creativity (e.g., Keri, 2009). For instance, Smalley, Loo, Yang, and Cantor (2005) proposed that, as atypical cerebral asymmetry (ACA) has been featured both in certain aspects of creativity and in a number of neuropsychiatric conditions, genetic risk factors underlying ACA may also be genetic enhancer factors for creativity (Smalley et al., 2005). However, the relationship between creativity and psychopathology is complex, differing between the groups of individuals in question, the nature of the illness, and the environmental factors involved (Glazer, 2009). As an illustration, Kinney, Richards, Lowing, LeBlanc, Zimbalist, and Harlan (2001) compared thirty-six index adult adoptees of biological parents with schizophrenia and thirty-six demographically matched control adoptees with no biological family history of psychiatric hospitalization. Those who did not have a clinical diagnosis of schizophrenia but had either schizotypal or schizoid personality disorder or multiple schizotypal signs in both groups (which other research has linked with a genetic liability for schizophrenia) had significantly higher creativity achievements (according to an interview-based measure) than other study participants. Reciprocally, it has been proposed that genes associated with affective disorders (e.g., mania, associated psychotic states, schizophrenia) serve as genetic reservoirs from which "genes for genius" are drawn (Akiskal & Akiskal, 2007).

Creativity: An Individual, Genetically Based Ability?

The search for the etiologies of individual differences in creativity remains challenging. Thus far, the mechanisms underlying the genetic bases of creativity—on the molecular and neurobiological levels—are still unclear, and the work on the specific gene action involved is inconclusive. As creativity results from the interactions of many cognitive and conative components, the genetic influences underlying creative ability could thus be informed through a better understanding of the genetic bases of these components. A major difficulty of reaching such an understanding is that the characteristics presumed to indicate creativity at the level of behavior may exceed 300 (Treffinger, 2009). As reviewed above, the genetic bases of some of the characteristics important for creativity have already been mapped out, including factors such as divergent thinking (Kéri, 2009) and sensation seeking (e.g., Ebstein et al., 1996; Koopmans, Boomsma, Heath, & Doornen, 1995). Unfortunately, though, as it is now generally accepted that all behavior has a genetic basis, heritability estimates are found to be only an initial—more suggestive than revelatory—indicator of the importance of genetic influences (Johnson, Penke, & Spinath, 2011a). In other words, heritability estimates validate a general belief that all behavior is inherited to some extent, yet these estimates have little explanatory power to support that belief, being characterized by underlying measurement error, statistical artifacts, epigenetic mechanisms, gene-environment interactions, and possibly other confounding factors yet to be discovered (Johnson, Penke, & Spinath, 2011b).

The Second Part of the Equation: The Reception of Creative Products

We now turn our attention from the individual creator to how the social world receives individuals' creative products, in order to understand the second part of the creativity phenomenon: the mechanisms that play a role in defining what is considered to be creative, and, therefore, lead to the adoption or rejection of new products. What are the underlying genetic bases of the social processes that define creativity, if there are any? First, we will delineate how creativity is a cultural phenomenon, as discussed in the literature. Then, we will present creativity's role in evolution, and the individual and social factors involved in creativity's reception. Finally, we will present considerations of how creativity, as a part of culture, may affect the genome, as outlined by theories of gene-culture coevolution and transmission.

Creativity as a Cultural Phenomenon

Centuries of philosophical thinking and several decades of individualistic psychological theorizing have located creativity in persons or products that in some way "stand apart" from their social background (Glăveanu, 2010). However, the human features involved in creativity—creative "ability"—represent only one side of the creativity coin. The effects of the cultural environment on creativity can also be profound (Lubart, 1990)—first, because it is within cultures that creative products are valued, but also because cultures affect the definition of creativity, the creative process, the direction in which creativity is channeled, and the degrees to which creativity is nurtured (Lubart, 1990). For Selby and colleagues (2005), it is clear that creative behavior is influenced by the "match" or "mismatch" between personality (i.e., individual factors) and environment, supporting the dictum "What is honored in a culture will be cultivated there."[1] Consistent with this idea, Csikszentmihalyi (1999) claims that the phenomenon of creativity is as much a cultural and social as it is a psychological event.

Yet, an essential aspect of creativity involves a creative contribution being considered valuable by an audience, recipient, or evaluator (Hempel & Sue-Chan, 2010). Nowadays, diverse creativity theories converge on the importance of environmental interactions with individual characteristics (Selby et al., 2005), and there is much evidence for the ways in which culture influences creators, what they produce, and how the products resulting from individual efforts are subject to the judgment of the social world. Notions of "adaptation," "appropriateness," or "value" are often part of the definition of creativity (e.g., Amabile, 1996; Barron, 1988; MacKinnon, 1962; Ochse, 1990; Sternberg & Lubart 1995), emphasizing that a creative contribution must be recognized as such by a culture or society. Stein (1953) elaborates that a creative contribution has to be accepted as tenable or useful or satisfying by a group at some point in time, emphasizing a second important aspect of culture in the recognition of creativity: the notion of timing. The literature abounds with examples of famous creative contributions that were not recognized by their culture in their time (e.g., Van Gogh, Bach). These examples illustrate the sociocultural relativity of creativity. If the product is "only" defined as original, the nonadaptive aspect makes it strange, bizarre, and consequently not creative (relative to a particular culture at a particular time). In other words, creative products are not defined as creative if they are not identified as new and valuable, appropriate, or adaptive. However, in the art domain, Lubart (1990) has suggested a possible "universality" in the judgments of aesthetic value by experts, even though many specific characteristics of creative phenomena depend on the cultural environment. What might be the origins of such "universal" judgments? Apart from being different from other products (i.e., new, original), how is a creative product recognized as being adaptive or valuable in a particular culture, or even universally? What are the cultural processes by which a new product is adopted and progressively becomes the "new norm"? And how, beyond cultural differences, are these processes genetically grounded (or not)?

Creativity in Evolution and Creativity as Evolution

Many authors have argued that aesthetic and creative experiences and abilities are part of human nature (e.g., Feist, 2007). For instance, Henri Bergson (1911) in *The Creative Evolution* describes the unique ability of humans to create artificial objects, especially tools to make tools, infinitely varying their making. Correspondingly, Witt (2003) argues that tool creation is a qualitatively distinct, and very rare, form of production in nature, relying on knowledge transmitted from generation to generation throughout the ontogeny of humans. The question is, how is this type of knowledge transmitted?

Before exploring the possible underlying mechanism of such phenotypic knowledge (applied to the "adoption" of novelty by a culture), it is worth noting that a growing body of empirical evidence supports the idea that human creativity and the aesthetic response has been shaped by evolutionary pressures (e.g., Barrow, 1995; Coss, 1968; Dissanayake, 1988, 2007; Feist, 2001; Miller, 2000). The historic (from prehistory on) occurrence of art, in some form or manifestation, in all cultures has led many to theorize on and investigate the adaptive or evolutionary aspects of creativity. Summarizing these theorizations, Dissanayake (2007) identified nine possible evolutionary uses of the arts (i.e., visual art, music, dance, and visual performance) for society: (1) as a way to acquire deeper knowledge of objects (Solso, 1994; Zeki, 1999) or to solve perceptual problems (Ramachandran & Hirstein, 1999); (2) as a medium for promoting selective attention and positive emotional responses (i.e., the positive effect of beauty) that lead to adaptive decisions and problem solving (Feist, 2007; Orians, 2001; Yusuf, 2009); (3) as a part of mate selection (Miller, 2000, 2001); (4) as symbolic gestures of commitment (as in art-filled religious ceremony; Irons, 2001); (5) as play or make-believe that serves as risk-free practice for later in life (Tooby & Cosmides, 2001); (6) to manipulate and control others, as propaganda (Aiken, 1998); (7) to enhance social cohesion, continuity, and cooperation, as in cultural ritual traditions (Boyd, 2005); (8) as a form of symbology to contribute to higher thinking and intelligence; and finally, (9) as a nonfunctional entity—something that solely offers aesthetic pleasure (Pinker, 1997, 2002). These hypotheses propose the possible functional roles of art that may form the basis for society's receptivity to art, the historical prevalence of art making and other forms of creativity, and the possible roots of creativity in human evolution.

Among other theorizations (see Dissanayake, 2007), Feist (2007) outlines an evolutionary theory of aesthetics grounded in natural and sexual selection. Natural selection focuses on solutions that solve survival problems, whereas sexual selection provides solutions that solve social and reproductive problems. Feist (2007) speculates that these two forces of selection (as evolutionary processes) have been important in shaping human creative potential and behavior over the millennia, but that each has shaped a distinct form of human creativity: the more "applied" forms of creativity (technology, engineering, and tool making) are probably more under natural selection pressures in that they have direct implications for one's surviving to reproductive age. By contrast, the more ornate and aesthetic forms of creativity (e.g., art, music, dance) are probably shaped more by sexual selection pressures, insofar as they implicitly signal an individual's genetic, physical, and mental fitness and are deemed attractive by members of the opposite sex (Feist, 2007; Miller, 2001). Following Feist's reasoning, it is possible to hypothesize that the human responses to these forms of creativity (i.e., the adoption of novelty, the aesthetic response) have been shaped according to the same evolutionary processes.

Feist's theory of aesthetics is supported by empirical research carried out with infants on the origins and uses of aesthetics (e.g., Kogan, 1994). Indeed, with respect to children's responses to creativity—either creating aesthetic objects or responding to them—Gardner (1982) has shown that all children exhibit a generalized pattern of development, suggesting that these responses are evolved domains of mind. For example, infant studies have shown that newborns have a preference for attractive over less attractive faces (Langlois, Ritter, Roggman, & Vaugh, 1991; Rubenstein, Kalakanis, & Langlois, 1999), consistently looking longer at the more attractive faces. Such studies of facial aesthetics have been conducted across ethnicities, and these do show agreement of what is considered an attractive female face across both ethnic and racial groups (Cunningham, 1991). Similarly, another set of studies on infant appreciation of melody (e.g., Krumhansl & Jusczyk, 1990; Trehub, 1987) also demonstrated early musical aesthetic response in human. For example, it has been found that six-month-old infants can detect mistunings in scales derived from their own native traditions as well as from unfamiliar traditions, while adults can more easily detect mistuning in the music of their own culture (Lynch, Eilers, Oller, & Urbano, 1990). Whether this can be explained adaptively or not, musicality (i.e., musical aesthetic response) appears to be a built-in feature of being human (Kogan, 1994). Other studies on specific musical abilities appear to confirm a possible genetic contribution. In a twin study carried out by Drayna and collaborators (Drayna, Manichaikul, de Lange, Snieder, & Spector, 2001), 136 monozygotic ("identical") twins and 148 dizygotic ("fraternal") twins were required to detect out-of-key notes in popular melodies. Performance was more similar between identical ($r = .79$) than between fraternal twins ($r = .46$), suggesting the importance of genetic influence in this particular facet of musical ability that could reflect the roots of some aspects of aesthetic judgment, possibly encoded in the genome (Kogan, 1994). Such evidence could thus explain the possible roots of the "universality" of aesthetic judgment, beyond cultural differences, as suggested by Lubart (1990).

Judging Creative Products: Individual and Social Processes and Their Underlying Genetic Bases

At the individual level, what are the factors that contribute to the judgment, appreciation, or adoption of new products, and how are these processes biologically and genetically grounded? The cultural roots of mental abilities can be mapped onto biological functions, since our perceptions, feelings, emotions, and judgments toward creative products are tied to physical structures (i.e., sensory organs and the brain) that are themselves dependent on the genome. Contrary to expectation, these issues are increasingly addressed by researchers in economic-related fields, including consumer psychology, cultural economics, or microeconomics. The adoption of novelty (or innovation-adoption) is, within these fields, a topic of major interest as it may explain change in consumption patterns (e.g., Ruprecht, 2005) and thus represent strong potential at a practical (i.e., commercial) level. Other topics of interest related to the reception of creative products from a consumer behavior perspective generally relate to the components of general consumer innovation-adoption (Wood & Swait, 2002), including the general "cultivation of taste" (McCain, 1979), optimal stimulation level (leading to the adoption of novelty), variety seeking, novelty seeking, exploratory tendencies, and information seeking (Hirschman, 1980; Raju, 1980; Steenkamp & Baumgartner, 1992; Venkatraman & Price, 1990). Openness to experience (e.g., Feist & Brady, 2004) and sensation seeking have also received major attention in innovation-adoption studies (e.g., Schweizer, 2006), while the genetic bases of these personality traits have also been explored as discussed above (e.g., Ebstein et al., 1996; Koopmans et al., 1995).

Beyond personality and other individual-related factors that contribute to individual differences in the perception and appreciation of novelty, some factors more anchored in evolutionary processes may also play a role, as emphasized through work in microeconomics. For example, Ruprecht (2005) presents an evolutionary approach to consumption theory, which highlights the role of consumer learning to explain the complex history of sweetener consumption—sweeteners being a good example of a "new" product that has been widely adopted in consumers' patterns. Building on the "continuity hypothesis" (Witt, 1993), which considers the evolution of culture to be based on biological evolution, Ruprecht (2005) argues that two evolutionary learning processes—reinforcement learning and social cognitive learning processes—explain the adoption of sweeteners by diverse cultures. The "continuity hypothesis" proposed by Witt (2003) suggests that human preferences result from learning processes that rely on certain universal "wants" that have been formed during human phylogenesis. These wants (including the need for air, nutrients, sexual activity) have a genetic basis and are considered to be "primary reinforcers" (Ruprecht, 2005), as conceptualized in classical conditioning. When these primary reinforcers are regularly paired with other items, the latter then obtain reinforcing potential and become secondary reinforcers. These secondary reinforcers are thus "learned" and act as rewarding experiences, and preferences of higher-order emerge (Ruprecht, 2005). During the individual learning history, a spectrum of "wants" or a "preference order" evolves (Witt, 2003). This reinforcement learning can also have a cultural aspect. Through reinforcement processes, preferences that are acquired by one generation are transferred to the next generation (e.g., parents tend to expose their children to goods that they like themselves). Ruprecht (2005) also identified the importance of social cognitive learning processes (the social version of reinforcement learning), underlying the role of consistency or "tightness" of individual choices with the prevailing social conventions and norms. Ruprecht (2005) illustrates this process in the context of the development of sweetener consumption, arguing that, for example, the consistency of beliefs with social norms is more crucial than proper scientific evidence for the alleged effects of sweeteners (given that the avoidance of obesity has become a normative imperative for certain groups of consumers).

Correspondingly, research shows that the "taste" for creative and artistic goods is not given only once for all time, but is dependent upon repeated exposure and experience (e.g., Bigand & Poulin-Charronnat, 2006). Berns, Capra, Moore, and Noussair (2010), for example, have observed the prevalence of conformism for the choices and appreciation of popular songs by teenagers, which could be interpreted in light of the sociocognitive learning process described by Ruprecht (2005). Berns and colleagues (2010) showed that the tendency to change one's evaluation of a song was correlated with activation only in the anterior insula, a region associated with physiological arousal, particularly to negative affective states. They interpret their results by suggesting an underlying mechanism whereby people are motivated to switch their choices in the direction of the consensus, through the anxiety generated by the mismatch between one's own preferences and others. This reinforces the importance of the role of consistency in individual choices with the prevailing social conventions and norms in innovation-adoption, as a social process. "Cultivation of taste" (e.g., McCain, 1979) may also be the reason for the existence of distinct preference traditions in different nations, regions, and population groups. This tendency of the population to fall into quite distinct and relatively homogeneous groups suggests multiple equilibria in the determination of tastes and demands (McCain, 1979). According to McCain (1979), these multiple equilibria may explain a sudden flowering of taste, the division of the population into "fan" in-groups and nonfan out-groups, seemingly arbitrary and mutable national traditions of drinking coffee or tea, and similar phenomena. As per recent mathematical models (Laland, Odling-Smee, & Myles, 2010), it is likely that this type of "niche" construction due to cultural processes could modify the selection on human genes.

Gene × Culture × Creativity Interaction

Social processes are thus part of the equation in the evaluation and adoption of novelty, even though these processes can be mapped onto biological and genetic roots. Genes and culture are thus partners in determining what creative products will be produced and valued (Kogan, 1994); and, as genetic features of humans may explain particular cultural events such as creative expression, production, and reception, it is also likely that reciprocally, cultural events may have influenced the evolution of the genome. In fact, a few works have attempted to better define the mechanisms by which genes and culture interact (e.g., Laland et al., 2010). For example, gene-culture coevolution studies, based on a theory first proposed by Lumsden and Wilson (1981), provide evidence that cultural practices, and not only large-scale environmental events, may influence the evolution of the genome. This theory attempts to establish a link between biological and cultural evolution, by which culture is shaped by biological constraints, and biological traits are simultaneously altered by the genetic evolution brought about by cultural innovations. More recent work building on these ideas (Laland et al., 2010) presents two plausible models of gene-culture interaction: (1) gene-culture coevolution and (2) niche-construction theory. The former generally builds on conventional population genetic theory, tracking the changes in allele and genotype frequencies in response to the more typically studied evolutionary processes such as genetic selection and random genetic drift, but also incorporating cultural transmission. The latter derives from evolutionary biology and is based on the capacity of organisms to select and modify natural selection in their environments, such as their use of particular plants to build nests modifying nutrient cycles, thus affecting evolutionary outcomes. Mathematical models show that niche construction due to cultural processes can modify selection on human genes (Laland et al., 2010). Findlay and Lumsden (1988) updated their gene-culture theory on the evolution of the creative mind; in their later theoretical model, the genotype, brain development, the cognitive phenotype (creative individual), and the sociocultural environment are interconnected to represent the multiple interactions and effects of discovery and innovation. Innovation is posited to affect not only the sociocultural and physical environment, but also the genetic composition of the next generation, either through natural selection or nonselective evolutionary mechanisms, that is, gene-culture transmission.

Conclusion

In this chapter, we have sketched out some of the less well-studied facets of creativity, particularly those that reflect creativity's dependence on a receptive environment, and we have explored the relationship of that dependence with genetic (biological and evolutionary) mechanisms. We have examined both addends of creativity's equation—that of the individual creator and that of the culture that receives the creative product—finding that both have notable and suggestive ties to the genome and therefore may be subject to the forces of evolution (while contributing to it). What does this mean for our understanding of creativity?

First, it is likely that many of the factors that contribute to creativity as an individual ability (e.g., novelty seeking, openness to experience), also contribute to the reception of creative products. Thus, much of the work on the genetic roots of creativity as an individual ability can help elucidate the "second part" of the creative equation: the genetic bases of the reception of creative products. Second, though creative advancements have been more typically credited to the "force of the creative individual," from the perspectives presented here, we can see that perhaps the evolutionary demand and need of a culture for new solutions have been equally forceful in making creativity happen. Conversely, we have seen that, by sociocultural mechanisms, creative products may also change the course of evolution and impact the genome. These mechanisms include social cognitive and "reinforcement" learning processes, which clarify how the consistency of beliefs with social norms (or predominant "taste") contributes to the adoption of novelty. Such mechanisms of the "standardization of tastes" can be mapped onto biological functions (how our perceptions, feelings, and judgments toward creative products are tied to biological structures) that are themselves dependent on the genome. In other words, it is important to see creativity not only as an individual "ability," but also as a cultural and time-specific phenomenon that is biologically grounded and has a social purpose.

With respect to future research on this topic, it is worth noting that many evolutionary approaches have speculated on how creativity and the reception of creative products may be a built-in feature of humans, and that many of these approaches have been heavily criticized as they tend to be highly theoretical. However, recent advancements in the research on the genetic bases of creativity tend to support these evolutionary hypotheses. Thus, further genetic studies examining these evolutionary hypotheses may discern the actual mechanisms of gene-culture transmission of creative ability, and for how creative products are received. Such research would contribute to the understanding of creativity as an essential feature of human adaptation and evolution (e.g., for the generation of new solutions to newly evolving problems). Indeed, in the creativity equation presented here, it is inescapably clear that wherever humans exist, creative productions will be made. There is *no* creativity without the social world, and there is no social world without the genetic forces that substantiate humans and humanity.

Acknowledgments

The preparation of this chapter was supported by funding from the National Institutes of Health, administered through grant RO1 DA01076; King Faisal University, Alhassa, Saudi Arabia, through grant R09517 (The Etiological Bases of Giftedness); and through the generous support of Karen Jensen Neff and Charlie Neff (The Aurora Project).

Note

1. Dictum attributed either to Plato (e.g., Selby et al., 2005) or to Aristotle (e.g., Torrance, 1998).

References

Aiken, N. E. (1998). *The biological origins of art.* Westport, CT: Praeger.

Akiskal, H. S., & Akiskal, K. K. (2007). In search of Aristotle: Temperament, human nature, melancholia, creativity, and eminence. *Journal of Affective Disorders, 100,* 1–6.

Amabile, T. M. (1996). *Creativity in context.* Boulder, CO: Westview.

Barbot, B., Besançon, M., & Lubart, T. I. (2011). Assessing creativity in the classroom. *Open Education Journal, 4*(1), 58–66.

Barron, F. (1988). Putting creativity to work. In R. J. Sternberg (Ed.), *The nature of creativity: Contemporary psychological perspectives* (pp. 77–98). New York: Cambridge University Press.

Barrow, J. D. (1995). *The artful universe*. Boston: Little, Brown.

Benjamin, J., Li, L., Patterson, C., Greenburg, B. D., Murphy, D. L., & Hamer, D. H. (1996). Population and familial association between the D4 dopamine receptor gene and measures of novelty seeking. *Nature Genetics, 12*, 81–84.

Bergson, H. (1911). *Creative evolution*. [*L'evolution creatice*, 1907.] (Mitchell, A., Trans.) New York: Henry Holt.

Berns, G., Capra, M., Moore, S., & Noussair, C. (2010). Neural mechanisms of the influence of popularity on adolescent ratings of music. *NeuroImage, 49*, 2687–2696.

Bigand, E., & Poulin-Charronnat, B. (2006). Are we experienced listeners? A review of the musical capacities that do not depend on formal musical training. *Cognition, 100*, 100–130.

Boyd, B. (2005). Evolutionary theories of art. In J. Gottschall & D. S. Wilson (Eds.), *Literature and the human animal* (pp. 147–176). Evanston, IL: Northwestern University Press.

Butcher, L. M., Meaburn, E., Knight, J., Sham, P. C., Schalkwyk, L. C., Craig, I. W., et al. (2005). SNPs, microarrays, and pooled DNA: Identification of four loci associated with mild mental impairment in a sample of 6,000 children. *Human Molecular Genetics, 14*, 1315–1325.

Buyske, S., Bates, M. E., Gharani, N., Matise, T. C., Tischfield, J. A., & Manowitz, P. (2006). Cognitive traits link to human chromosomal regions. *Behavior Genetics, 36*, 65–76.

Chavez-Eakle, R. A., Graff-Guerrero, A., García-Reyna, J.-C., Vaugier, V., & Cruz-Fuentes, C. (2007). Cerebral blood flow associated with creative performance: A comparative study. *NeuroImage, 38*, 519–528.

Cloninger, C. R. (1994). *The temperament and character inventory (TCI): A guide to its development and use*. St. Louis, MO: Washington University, Center for the Psychobiology of Personality.

Cloninger, C. R., Przybeck, T. R., & Svrakic, D. M. (1991). The tridimensional personality questionnaire—United States normative data. *Psychological Reports, 69*, 1047–1057.

Coss, R. G. (1968). The ethological command in art. *Leonardo, 1*, 273–287.

Csikszentmihalyi, M. (1997). *Creativity: Flow and the psychology of discovery and invention*. New York: Harper Collins.

Csikszentmihalyi, M. (1999). Implications for a systems perspective for the study of creativity. In R. J. Sternberg (Ed.), *Handbook of creativity* (pp. 313–335). New York: Cambridge University Press.

Cunningham, M. R. (1991). A psycho-evolutionary, multiple-motive interpretation of physical attractiveness. Paper presented at the 99th Convention of the American Psychological Association.

Curtis, D. (2004). Re-analysis of collaborative study on the genetics of alcoholism pedigrees suggests the presence of loci influencing novelty-seeking near D12S391 and D17S1299. *Psychiatric Genetics, 14,* 151–155.

Dacey, J. S. (1989). Discriminating characteristics of the families of highly creative adolescents. *Journal of Creative Behavior, 23,* 263–271.

Davies, G., Tenesa, A., Payton, A., Yang, J., Harris, S. E., Liewald, D., et al. (2011). Genome-wide association studies establish that human intelligence is highly heritable and polygenic. *Molecular Psychiatry, 16,* 996–1005.

Deary, I. J., Johnson, W., & Houlihan, L. M. (2009). Genetic foundations of human intelligence. *Human Genetics, 126,* 215–232.

Deary, I. J., Penke, L., & Johnson, W. (2010). The neuroscience of human intelligence differences. *Nature Reviews: Neuroscience, 11,* 201–211.

Deary, I. J., Spinath, F. M., & Bates, T. C. (2006). Genetics of intelligence. *European Journal of Human Genetics, 14,* 690–700.

Devlin, B., Daniels, M., & Roeder, K. (1997). The heritability of IQ. *Nature, 388,* 469–471.

Dick, D. M., Aliev, F., Bierut, L. J., Goate, A., Rice, J., Hinrichs, A., et al. (2006). Linkage analysis of IQ in the collaborative study on the genetics of alcoholism (COGA) sample. *Behavior Genetics, 36,* 77–86.

Dissanayake, E. (1988). *What is art for?* Seattle, WA: Washington University Press.

Dissanayake, E. (2007). What art is and what art does: An overview of contemporary evolutionary hypotheses. In C. Martindale (Ed.), *Evolutionary and neurocognitive approaches to aesthetics, creativity, and the arts* (pp. 1–14). Amityville, NY: Baywood Publishing.

Drayna, D., Manichaikul, A., de Lange, M., Snieder, H., & Spector, T. (2001). Genetic correlates of musical pitch recognition in humans. *Science, 291,* 1969–1972.

Ebstein, R. P., Nemanov, L., Klotz, I., Gritsenko, I., & Belmaker, R. H. (1997). Additional evidence for an association between the dopamine D4 receptor (DRD4) exon

III repeat polymorphism and the human personality trait of novelty seeking. *Molecular Psychiatry, 2,* 472–477.

Ebstein, R. P., Novick, O., Umansky, R., Priel, B., Osher, Y., Blaine, D., et al. (1996). Dopamine D4 receptor (DRD4) exon III polymorphism associated with the human personality trait of novelty seeking. *Nature Genetics, 12,* 78–80.

Estes, Z., & Ward, T. B. (2002). The emergence of novel attributes in concept modification. *Creativity Research Journal, 14,* 149–156.

Eysenck, H. J. (1983). The roots of creativity: Cognitive ability or personality trait? *Roeper Review, 5,* 10–12.

Feist, G. J. (2001). Natural and sexual selection in the evolution of creativity. *Bulletin of Psychology and the Arts, 2,* 11–16.

Feist, G. J. (2007). An evolutionary model of artistic and musical creativity. In C. Martindale, P. Locher, & V. M. Petrov (Eds.), *Evolutionary and neurocognitive approaches to aesthetics, creativity, and the arts* (pp. 15–30). Amityville, NY: Baywood Publishing.

Feist, G. J., & Brady, T. R. (2004). Openness to experience, non-conformity, and the preference for abstract art. *Empirical Studies of the Arts, 22,* 77–89.

Findlay, C. S., & Lumsden, C. J. (1988). The creative mind: Toward an evolutionary theory of discovery and innovation. *Journal of Social and Biological Systems, 11,* 3–55.

Folley, B. S., Doop, M. L., & Park, S. (2003). Psychoses and creativity: is the missing link a biological mechanism related to phospholipids turnover? *Prostaglandins, Leukotrienes, and Essential Fatty Acids, 69*(6), 467–476.

Gardner, H. (1982). *Art, mind, and brain: A cognitive approach to creativity.* New York: Basic Books.

Glaveanu, V.-P. (2010). Principles for a cultural psychology of creativity. *Culture and Psychology, 16,* 147–163.

Glazer, E. (2009). Rephrasing the madness and creativity debate: What is the nature of the creativity construct? *Personality and Individual Differences, 46,* 755–764.

Grigorenko, E. L., LaBuda, M. C., & Carter, A. S. (1992). Similarity in general cognitive ability, creativity, and cognitive style in a sample of adolescent Russian twins. *Acta Geneticae Medicae et Gemellologiae, 41,* 65–72.

Guilford, J. P. (1950). Creativity. *American Psychologist, 5,* 444–454.

Guilford, J. P. (1987). Creativity research: Past, present, and future. In S. G. Isaksen (Ed.), *Frontiers of creativity research: Beyond the basics* (pp. 61–85). Buffalo, NY: Bearly Limited.

Hempel, P. S., & Sue-Chan, C. (2010). Culture and the assessment of creativity. *Management and Organization Review, 6,* 415–435.

Hirschman, E. C. (1980). Innovativeness, novelty seeking, and consumer creativity. *Journal of Consumer Research, 7,* 283–295.

Irons, W. (2001). Religion as a hard-to-fake sign of commitment. In R. Nesse (Ed.), *Evolution and the capacity for commitment.* New York: Russell Sage.

Johnson, W., Penke, L., & Spinath, F. M. (2011a). Heritability in the era of molecular genetics: Some thoughts for understanding genetic influences on behavioural traits. *European Journal of Personality, 25,* 254–266.

Johnson, W., Penke, L., & Spinath, F. M. (2011b). Understanding heritability: What it is and what it is not. *European Journal of Personality, 25,* 287–294.

Karlsson, J. L. (1970). Genetic association of giftedness and creativity with schizophrenia. *Hereditas, 66,* 177–182.

Kaufman, A. B., Kornilov, S. A., Bristol, A. S., Tan, M., & Grigorenko, E. L. (2010). The neurobiological foundation of creative cognition. In J. C. Kaufman & R. J. Sternberg (Eds.), *The Cambridge handbook of creativity* (pp. 216–232). New York: Cambridge University Press.

Kéri, S. (2009). Genes for psychosis and creativity: A promoter polymorphism of the *neuregulin 1* gene is related to creativity in people with high intellectual achievement. *Psychological Science, 20,* 1070–1073.

Kinney, D. K., Richards, R., Lowing, P. A., LeBlanc, D., Zimbalist, M. E., & Harlan, P. (2001). Creativity in offspring of schizophrenic and control parents: An adoption study. *Creativity Research Journal, 13,* 17–25.

Kogan, N. (1994). On aesthetics and its origins: Some psychobiological and evolutionary considerations. *Social Research, 61,* 139–165.

Koopmans, J., Boomsma, D., Heath, A. C., & Doornen, L. (1995). A multivariate genetic analysis of sensation seeking. *Behavior Genetics, 25,* 349–356.

Krumhansl, C. L., & Jusczyk, P. W. (1990). Infants' perception of phrase structure in music. *Psychological Science, 1,* 70–73.

Laland, K. N., Odling-Smee, J., & Myles, S. (2010). How culture shaped the human genome: bringing genetics and the human sciences together. *National Review, 11,* 137–148.

Langlois, J. H., Ritter, J. M., Roggman, L. A., & Vaugh, L. S. (1991). Facial diversity and infant preferences for attractive faces. *Developmental Psychology, 27,* 79–84.

Lerman, C., Audrain, J., Main, D., Bowman, E. D., Lockshin, B., Boyd, N. R., & Shields, P. G. (1999). Evidence suggesting the role of specific genetic factors in cigarette smoking. *Health Psychology, 18,* 14–20.

Lubart, T. I. (1990). Creativity and cross-cultural variation. *International Journal of Psychology, 25,* 39.

Luciano, M., Wright, M. J., Duffy, D. L., Wainwright, M. A., Zhu, G., Evans, D. M., et al. (2006). Genome-wide scan of IQ finds significant linkage to a quantitative trait locus on 2q. *Behavior Genetics, 36,* 45–55.

Lumsden, C. J., & Wilson, E. O. (1981). *Genes, mind, and culture.* Cambridge, MA: Harvard University Press.

Lynch, M. P., Eilers, R. E., Oller, D. K., & Urbano, R. C. (1990). Innateness, experience, and music perception. *Psychological Science, 1,* 272–276.

MacKinnon, D. W. (1962). The nature and nurture of creative talent. *American Psychologist, 17,* 484–495.

Mandelman, S. D., & Grigorenko, E. L. (2011). Intelligence: Genes, environments, and their interactions. In R. J. Sternberg & S. B. Kaufman (Eds.), *The Cambridge handbook of intelligence* (pp. 85–106). New York: Cambridge University Press.

McCain, R. A. (1979). Reflections on the cultivation of taste. *Journal of Cultural Economics, 3,* 30–52.

McCrae, R. R. (1987). Creativity, divergent thinking, and openness to experience. *Journal of Personality and Social Psychology, 52,* 1258–1263.

Miller, G. F. (2000). Evolution of human music through sexual selection. In N. L. Wallin, B. Merker, & S. Brown (Eds.), *The origins of music* (pp. 329–360). Cambridge, MA: MIT Press.

Miller, G. F. (2001). Aesthetic fitness: How sexual selection shaped artistic virtuosity as a fitness indicator and aesthetic preference as mate choice criteria. *Bulletin of Psychology and the Arts, 2,* 20–25.

Munafo, M. R., Yalcin, B., Willis-Owen, S. A., & Flint, J. (2008). Association of the dopamine D4 receptor (DRD4) gene and approach-related personality traits: Meta-analysis and new data. *Biological Psychiatry, 63,* 197–206.

Ochse, R. (1990). *Before the gates of excellence: The determinants of creative genius.* New York: Cambridge University Press.

Orians, G. H. (2001). An evolutionary perspective on aesthetics. *Bulletin of Psychology and the Arts, 2,* 25–29.

Patrick, C. L. (2000). Genetic and environmental influences on the development of cognitive abilities: Evidence from the field of developmental behavior genetics. *Journal of School Psychology, 38,* 79–108.

Pinker, S. (1997). *How the mind works.* New York: Norton Books.

Pinker, S. (2002). *The blank slate: The modern denial of human nature.* New York: Viking Penguin.

Plomin, R., & Spinath, F. M. (2004). Intelligence: Genetics, genes, and genomics. *Journal of Personality and Social Psychology, 86,* 112–129.

Posthuma, D., Luciano, M., Geus, E. J., Wright, M. J., Slagboom, P. E., Montgomery, G. W., et al. (2005). A genomewide scan for intelligence identifies quantitative trait loci on 2q and 6p. *American Journal of Human Genetics, 77*, 318–326.

Prolo, P., & Licinio, J. (2002). DRDF and novelty-seeing. In J. Benjamin, R. P. Ebstein, & R. H. Belmaker (Eds.), *Molecular genetics and the human personality*. Washington, D.C.: American Psychiatry Publishing.

Raju, P. S. (1980). Optimum stimulation level: Its relationship to personality, demographics, and exploratory behavior. *Journal of Consumer Research, 7*, 272–282.

Ramachandran, V. S., & Hirstein, W. (1999). The science of art: A neurological theory of aesthetic experience. *Journal of Consciousness Studies, 6*, 15–51.

Reuter, S., Roth, S., Holve, K., & Hennig, J. (2006). Identification of first candidates genes for creativity: A pilot study. *Brain Research, 1069*, 190–197.

Reznikoff, M., Domino, G., Bridges, C., & Honeyman, M. (1973). Creative abilities in identical and fraternal twins. *Behavior Genetics, 3*, 365–377.

Richards, R., & Kinney, D. K. (2000). Mood disorders and what we all can learn. *Bulletin of Psychology and the Arts, 1*, 44–46.

Roe, B. E., Tilley, M. R., Gu, H. H., Beversdorf, D. Q., Sadee, W., Haab, T. C., et al. (2009). Financial and psychological risk attitudes associated with two single nucleotide polymorphisms in the nicotine receptor (CHRNA4) gene. *PLoS ONE, 4*, e6704.

Rubenstein, A. J., Kalakanis, L., & Langlois, J. H. (1999). Infant preferences for attractive faces: A cognitive explanation. *Developmental Psychology, 35*, 848–855.

Runco, M. A. (2004). Creativity. *Annual Review of Psychology, 55*, 657–687.

Ruprecht, W. (2005). The historical development of the consumption of sweeteners: A learning approach. *Journal of Evolutionary Economics, 15*, 247–272.

Scheinfeld, A. (1973). *Twins and supertwins*. Oxford: Penguin.

Schweizer, T. S. (2006). The psychology of novelty-seeking, creativity and innovation: Neurocognitive aspects within a work-psychological perspective. *Creativity and Innovation Management, 15*, 164–172.

Selby, E. C., Shaw, E. J., & Houtz, J. C. (2005). The creative personality. *Gifted Child Quarterly, 49*, 300–314.

Smalley, S. L., Loo, S. K., Yang, M. H., & Cantor, R. M. (2005). Toward localizing genes underlying cerebral asymmetry and mental health. *American Journal of Medical Genetics. Part B, Neuropsychiatric Genetics, 135B*, 79–84.

Solso, R. (1994). *Cognition and the visual arts*. Cambridge, MA: MIT Press.

Steenkamp, J. E. M., & Baumgartner, H. (1992). The role of optimum stimulation level in exploratory consumer behavior. *Journal of Consumer Research, 19*, 434–448.

Stein, M. I. (1953). Creativity and culture. *Journal of Psychology, 36,* 311–322.

Sternberg, R. J. (2006). The nature of creativity. *Creativity Research Journal, 18,* 87–98.

Sternberg, R. J., & Lubart, T. I. (1995). *Defying the crowd: Cultivating creativity in a culture of conformity.* New York: Free Press.

Tooby, J., & Cosmides, L. (2001). Does beauty build adapted minds? Toward an evolutionary theory of aesthetics, fiction, and the arts. *SubStance, 30,* 6–27.

Torrance, E. P. (1998). Reflection on emerging insights on the educational psychology of creativity. In J. Houtz (Ed.), *The educational psychology of creativity* (pp. 283–286). Cresskill, NJ: Hampton Press.

Treffinger, D. J. (1996). *Creativity, creative thinking, and critical thinking: In search of definitions.* Sarasota, FL: Center for Creative Learning.

Treffinger, D. J. (2009). Myth 5: Creativity is too difficult to measure. *Gifted Child Quarterly, 53,* 245–247.

Trehub, S. E. (1987). Infants' perception of musical patterns. *Perception & Psychophysics, 41,* 635–641.

Venkatraman, M. P., & Price, L. L. (1990). Differentiating between cognitive and sensory innovativeness: Concepts, measurement, and implications. *Journal of Business Research, 20,* 293–315.

Vernon, P. E. (1989). The nature-nurture problem in creativity. In J. A. Glover, R. R. Ronning, & C. R. Reynolds (Eds.), *Handbook of creativity: Perspectives on individual differences* (pp. 93–110). New York: Plenum Press.

Verweij, K. J. H., Zietsch, B. P., Medland, S. E., Gordon, S. D., Benyamin, B., Nyholt, D. R., et al. (2010). A genome-wide association study of Cloninger's temperament scales: Implications for the evolutionary genetics of personality. *Biological Psychology, 85,* 306–317.

Volf, N. V., Kulikov, A. V., Bortsova, C. U., & Popova, N. K. (2009). Association of verbal and figural creative achievement with polymorphisms in the human serotonin transporter gene. *Neuroscience Letters, 463,* 154–157.

Vygotsky, L. (1971). *The psychology of art.* Cambridge, MA: MIT Press.

Wainwright, M. A., Wright, M. J., Luciano, M., Montgomery, G. W., Geffen, G. M., & Martin, N. G. (2006). A linkage study of academic skills defined by the Queensland Core Skills Test. *Behavior Genetics, 36,* 56–64.

Witt, U. (1993). *Evolutionary economics.* Aldershot: Edward Elgar.

Witt, U. (2003). Evolutionary economics and the extension of evolution to the economy. In U. Witt (Ed.), *The evolving economy* (pp. 3–34). Cheltenham: Elgar.

Wood, S. L., & Swait, J. (2002). Psychological indicators of innovation adoption: Cross-classification based on need for cognition and need for change. *Journal of Consumer Psychology, 12,* 1–13.

Yusuf, S. (2009). From creativity to innovation. *Technology in Society, 31,* 1–8.

Zeki, S. (1999). *Inner vision: An exploration of art and the brain.* Oxford: Oxford University Press.

Zuckerman, M. (1994). *Behavioral expressions and bio-social expressions of sensation seeking.* Cambridge: Cambridge University Press.

5 Creativity and Talent: Etiology of Familial Clustering

Marleen H. M. de Moor, Mark Patrick Roeling, and Dorret I. Boomsma

The question of whether an individual has inherited a particular skill or talent or whether skills and talent are acquired through practice and training has long been a topic of debate. On this question, Sir Francis Galton (1822–1911), a cousin of Darwin, stated: "I have no patience with the hypothesis occasionally expressed, and often implied, especially in tales written to teach children to be good, that babies are born pretty much alike, and that the sole agencies in creating differences between boy and boy, and man and man, are steady application and moral effort." Galton undertook several efforts to address the issue in an empirical manner. In his books *Hereditary Genius: An Inquiry into Its Laws and Consequences* (1869) and *English Men of Science: Their Nature and Nurture* (1874), he investigated whether talent and giftedness possibly descend through ancestry. Through an inquiry into the relationships of eminent Englishmen, Galton observed high interrelatedness. Those included in his study more frequently had eminent fathers than grandfathers, and eminent sons than grandsons. Consequently, in the first- or second-degree relationship the degree of resemblance was higher than in third-degree relationships.

Familial clustering has also been described in various case studies of legendary families. For example, the Bach family showed a remarkable concentration of musical talent, and the Amati family introduced the violin into Italy. However, familial does not equal genetic (Martin, Boomsma, & Machin, 1997), and early exposure to favorable environments may explain the clustering that has been observed in pedigrees such as that of the Bach family. To investigate whether social advantages rather than biology underlie familial clustering, Galton introduced an early variant of the "adoption design," in which he used the adopted sons of Popes and their nephews as a control group.

Assessing the influence of genetics and environment is essential for our understanding of the etiology of individual differences in behavior and complex traits, including talents. In addition to the adoption design, Galton pioneered several techniques that have become critical methods in the field of behavior genetics. These methods include the use of twins and families to estimate the impact of heredity on individual differences. In behavioral genetics, the central quest is to find out to what extent individual differences in a measured trait (phenotype) arise from differences in genotype and differences in environment (i.e., the sum of all nongenetic influences). The variance in a phenotype is decomposed into components due to the effects of polymorphic genes at all locations across the genome, and those due to the effects of environment. Studies where this decomposition of variance is performed often result in an estimated measure called "heritability." Heritability is the proportion of phenotypic variance that can be accounted for by differences between individuals. Genetic influences are sometimes portioned into additive and nonadditive effects of alleles (an allele is a particular variant of a gene), while environmental effects are portioned into the effects of common environment shared by individuals growing up in the same family and the effects of nonshared environment (Plomin, DeFries, McClearn, & McGuffin, 2008).

The perception of talents and abilities as innate versus as products of environmental factors can have a large societal impact. A consequence of the belief that innate talent is a precondition for high achievement may be that young people who are not identified as having the talent could be more likely to be denied the help and encouragement they would need to attain high levels of competence (Howe, Davidson, & Sloboda, 1998). Moreover, insight into the etiology of traits, whether genetic, environmental, or both, potentially influences the definition and operationalization of that trait. In Galton's work, talent was operationalized in a rather unique manner. It was defined according to a person's enduring reputation, called *eminence*. By this, Galton (1869, p. 33) referred to "the opinion of contemporaries, revised by posterity … the reputation of a leader of opinion, of an originator, of a man to whom the world deliberately acknowledges itself largely indebted." In other works, *talent* has often been used interchangeably with the term *genius*. Successful composers, artists, poets, writers, and scientists began to be called geniuses and achieved fame for the realization of their special abilities. However, large variety exists within the term genius. Those who are highly intelligent (e.g., Einstein), highly eminent (e.g., Nobel Prize winners), or highly creative (e.g., Napoleon Bonaparte) are all referred to as geniuses (Simonton, 1999). In their review, Howe et al. (1998) assign five properties to talent: (1) it originates in genetically transmitted structures and hence is at least partly innate; (2) its full effects may not be evident at an early stage, but there will be some advance indications, allowing trained people to identify the presence of talent before exceptional levels of mature performance have been demonstrated; (3) these early indications of talent provide a basis for predicting who is likely to excel; (4) only a minority are talented, for if all children were, then there would be no way to predict or explain differential success; and finally, (5) talents are relatively domain specific (Howe et al., 1998, p. 399).

In this chapter, we aim to extend our prior work on talents (Vinkhuyzen, van der Sluis, Posthuma, and Boomsma, 2009) in which we studied the contribution of genetic factors to a series of talents in a general population sample of adolescent and young adult Dutch twins. The talents included music, arts, chess, writing, memory, knowledge, languages, and mathematics. Participants were asked to rate themselves on each of these domains. Other studies have also focused on the genetic architecture of some of these talents (Howe et al., 1998; Lubinsky et al., 2006; McGue et al., 1993; Ruthsatz et al., 2008). We briefly review the main findings for each of these talents below.

Individual differences in musical abilities have been attributed to genetic factors by Fuller and Thompson (1978). Heritability estimates range from 10 percent for nonschool musical performance to 71 percent for vocal performance (Coon & Carey, 1989). Pitch recognition is heritable for 70 to 80 percent (Drayna, Manichaikul, de Lange, Snieder, & Spector, 2001). Absolute pitch has also been found to aggregate in families (Baharloo et al., 2000).

Regarding chess, the nature-nurture debate was intensified by the Hungarian psychologist and chess teacher László Polgár. In his book *Bring Up Genius!* he emphasized the importance of specialized training (Polgár, 1989). He argued that by fully tapping into the potential of children, one could become capable of simply reading the chessboard and making decisions based on intuition. This belief was confirmed by a study of De Bruin and colleagues (2008) where most of the variation in performance was accounted for by deliberate practice. However, the question whether extensive practice reflects a genetic disposition to enjoy and benefit from playing and practicing chess remains unanswered.

Concerning the ability to learn and acquire a new language as well as impairments in language development, several studies have addressed the role of genetic factors and high heritability estimates (Dale et al., 1998; Newbury et al., 2005; Spinath et al., 2004; Stromswold, 2001; Viding et al., 2003). Recently, a twin study in young adolescents examined the heritability of second-language acquisition (Dale et al., 2010) and suggested a substantial heritability of 67 percent. The remaining variance was explained by shared environmental influences (13 percent) and unique environmental influences (20 percent). The authors mentioned that the influence of genetics in the ability to speak multiple foreign languages is higher than estimates in first-language acquisition, and that large overlap exists between the genes for first- and second-language acquisition (Dale et al., 2010).

Studies into the decomposition of the variance in what is called "general knowledge" derives, for example, from the Information subtest of the WAIS-III (1997). Heritability for the information subtest is estimated at 75 percent (Rijsdijk et al., 2002). Regarding memory function, consensus exists that genetic factors explain around 50 percent of individual differences (Bouchard, 1998; Finkel et al., 1995).

Mathematics has been studied from both biological and genetic perspectives. The biological basis for mathematical talent was explored, for example, in a study on sex differences in mathematics (Benbow & Lubinski, 1993). Hormonal influences, medical and bodily conditions, and specific brain activations are associated with mathematical achievement. Twin studies have reported heritability estimates ranging from 19 up to 90 percent (Alarcon et al., 2000; Thompson et al., 1991; Wijsman et al., 2004).

Overall, the impression from earlier studies is that both genetics and the environment contribute to variation in talents. Extending the classical twin design to a multigeneration design has several advantages, including a direct assessment of and test for the genetic and cultural transmission from one generation to the next. In addition, data from spouses (the parents of twins) provide information on nonrandom mating.

In this chapter, we include information on self-reported talents from twins and from their parents. The inclusion of parents in a parent-offspring model offers the possibility to test for several hypotheses regarding the genetic architecture of complex traits (e.g., Eaves, 1972; Martin, Boomsma, & Machin, 1997; Maes et al., 2009; Fulker, 1989). One of those hypotheses considers the relationship between parental phenotypes and those of their children and addresses the question of cultural versus genetic inheritance. In addition, having data on spouses offers the possibility to study assortative or nonrandom mating. This type of mating refers to the phenomenon where the phenotypes of the parents are correlated (Eaves, Last, Martin, & Jinks, 1977), for example, men with a particular talent may more often tend to marry women who also are talented. The consequences of assortative mating depend on the mechanism that leads to assortment, including social homogamy, social interaction, and phenotypic assortment (Heath & Eaves, 1985; Reynolds et al., 2006; van Leeuwen et al., 2008). Social homogamy refers to the phenomenon where individuals coming from similar social backgrounds are more likely to meet and marry each other. Social interaction occurs when spouses mutually influence each other because they spend time together. Phenotypic assortment refers to the phenomenon where individuals select each other based on their phenotype. If assortment is phenotypic, this can have important consequences for the genetic architecture of traits because it increases the additive genetic variance in the offspring generation and the covariance between additive genetic factors in first-degree relatives (Eaves et al., 1989; van Grootheest et al., 2008).

The combination of parental data and data from monozygotic (MZ) and dizygotic (DZ) twins makes for a strong design to establish and screen for genetic influences in behavioral dimensions, such as talents. The different degree of genetic relatedness of MZ and DZ twins is used to identify the relative contribution of genes and environment. MZ twins derive from one fertilized egg (one zygote) and are genetically (nearly) identical, whereas DZ twins derive from two different fertilized eggs (two zygotes) and share on average half of their segregating genes, just like nontwin brothers and sisters. If the resemblance for a particular trait is larger in MZ than in DZ twins, this strongly suggests that their larger genetic resemblance is important. From the observed phenotypic differences between twins, the heritability can be estimated, which, if greater than zero, implies that genetic factors influence a trait (Boomsma et al., 2002).

When the shared environment is likely to have some influence on a trait, the parents-twins model provides the opportunity to estimate the influence of the phenotypes of the mother and father on the shared environment of the offspring. This implies that it is possible to partition the shared environmental effects found in talent among offspring into the effects of the parental phenotype on offspring behavior (vertical cultural transmission), environmental effects that are shared among offspring but not shared with parents (horizontal cultural transmission), and the effects of assortative mating. These effects can be separated by simultaneously modeling the correlations between parents, between parents and their offspring, and between offspring. When the spouse correlation is significant, part of the shared environmental effects may be explained by assortative mating. If the parent-offspring correlations are larger than what would be expected under genetic transmission alone, this could imply that the shared environmental effects found in offspring are the results of vertical cultural transmission. A lower parent-offspring correlation compared with DZ twin and sibling correlations can indicate that part of the environmental effects is shared between twins and siblings only, that part of the genetic factors acts in a dominant manner, or that different genetic factors affect talent in the two generations.

Netherlands Twin Registry (NTR) Study

The data for this study were collected in 1991 through a survey sent out by the Netherlands Twin Registry (NTR) established in 1987 (Boomsma et al., 2006). Adolescent and young adult twins and their parents were recruited through city council registrations and contacted by mail and invited to complete a survey (for details, see Koopmans, van Doornen, & Boomsma, 1994; Willemsen, Posthuma, & Boomsma, 2005). Data from 3,331 twins and 2,995 parents from 1,693 families were available. The twin sample consisted of 557 monozygotic males (MZM), 473 dizygotic males (DZM), 759 monozygotic females (MZF), 590 dizygotic females (DZF), 952 dizygotic opposite sex (DOS), and 1,407 fathers and 1,588 mothers. The mean age of the twins was 17.73 years (*SD* = 2.37, range = 12–50 years). The mean age for the parents was 46.6 years (*SD* = 5.4, range = 33–71 years).

The talents were selected from the Talent Inventory developed by McGue et al. (1993), and included self-report data on music, arts, chess, writing, memory, knowledge, language, and mathematics. Participants were asked to rank their own competence, in comparison to the general population, on an ordinal four-point scale representing different categories. The first category represents people who classify themselves as less competent than most people. The second category represents the average (as competent as most people), the third category the above average (more competent than most people), and the fourth category represents people who classify themselves at the top end, that is, as being exceptionally skilled.

Talent for Music refers to singing or playing one or more instruments. A talent for Arts reflects artistic and creative activities (e.g., painting, acting). Chess refers to the ability to play games like chess, backgammon, and mah-jong. Talent for Writing is defined as creative writing (e.g., letters, manuscripts, books). Memory reflects general mnemonic skills (e.g., events, numbers, and facts). Knowledge refers to general and specific knowledge of facts. Language is defined as the ability to speak one or more foreign languages. Finally, talent for Mathematics refers to mathematical and numerical ability.

The analysis of the categorical data was based on a "liability" model (Falconer & Mackay, 1996; Neale & Cardon, 1992). In this model, a categorical variable is assumed to reflect an imprecise measurement of an underlying normal distribution of liability. A threshold divides the population into two or more groups (categories) of people. For the study of talent—which was measured on a four-point scale—categories 3 and 4 were merged, and scores on the liability distribution could fall into three categories that were defined by two thresholds. Since the liability is a theoretical construct, its scale is arbitrary. The liability was assumed to be standard normally distributed with zero mean and unit variance. Correlations among family members on the liability scale and thresholds were estimated. Thresholds were allowed to vary as a function of sex (0 for male and 1 for female) and age (standardized). Age coefficients were similar for parents and for threshold 1 or 2. All analyses were conducted using structural equation modeling in Mx (Neale, Boker, Xie, & Maes, 2003).

Figures 5.1 to 5.8 show the prevalence of each talent in the adolescent boys and girls and in their fathers and mothers. In general, the proportion of individuals with exceptional talent was small. The percentage of individuals falling into the exceptionally talented category is typically less than 10 percent, except for talent in Languages. The proportion of individuals with above average but not exceptional aptitude is somewhat larger (ranging between 3 and 40 percent). When the two above average categories are combined, the percentage of individuals with above average talent ranges from 4.6 percent for Mathematics in mothers, to 42.5 percent for Memory

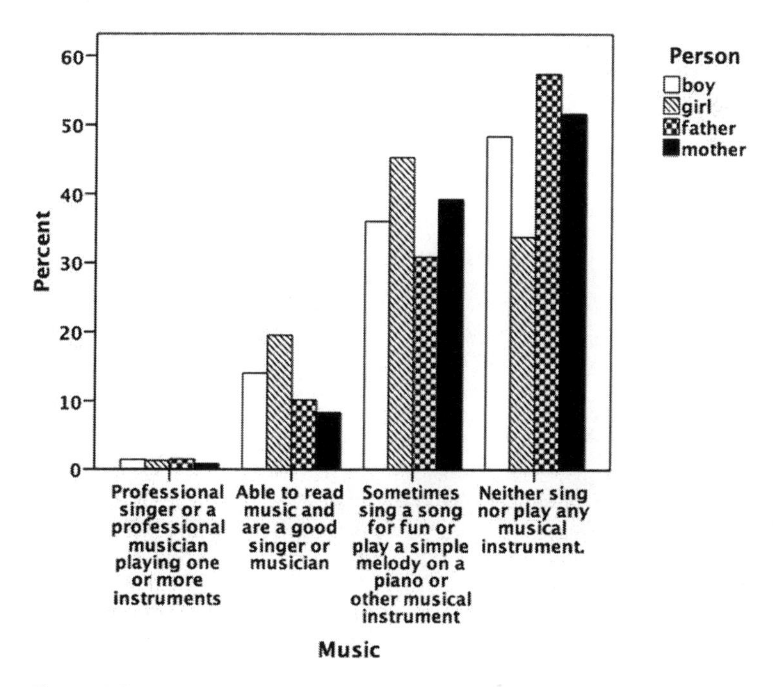

Figure 5.1
Prevalence of talent for music.

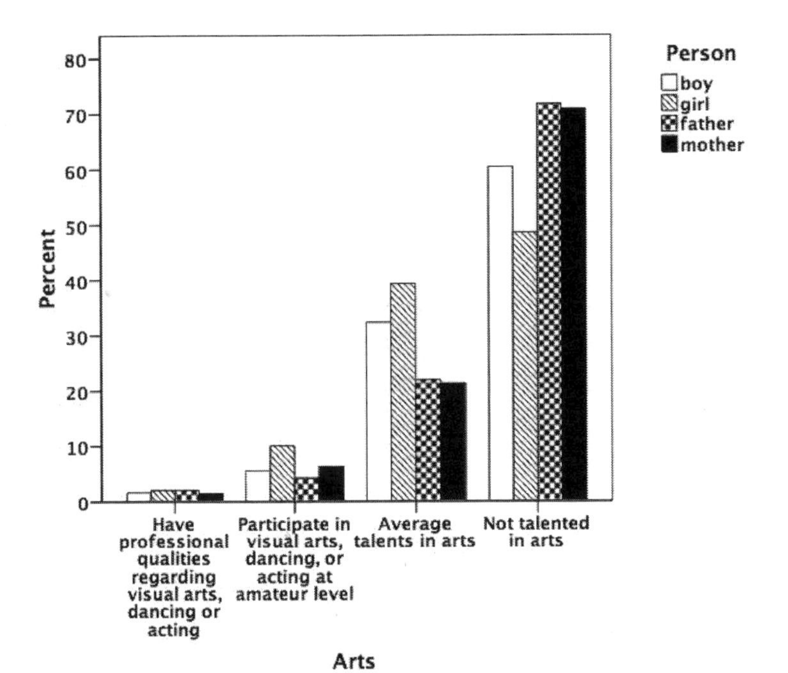

Figure 5.2
Prevalence of talent for arts.

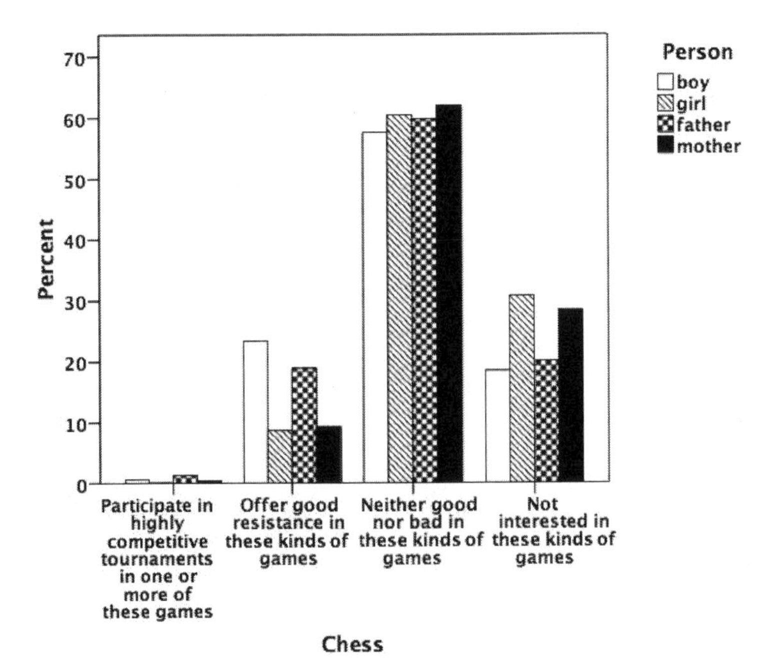

Figure 5.3
Prevalence of talent for chess.

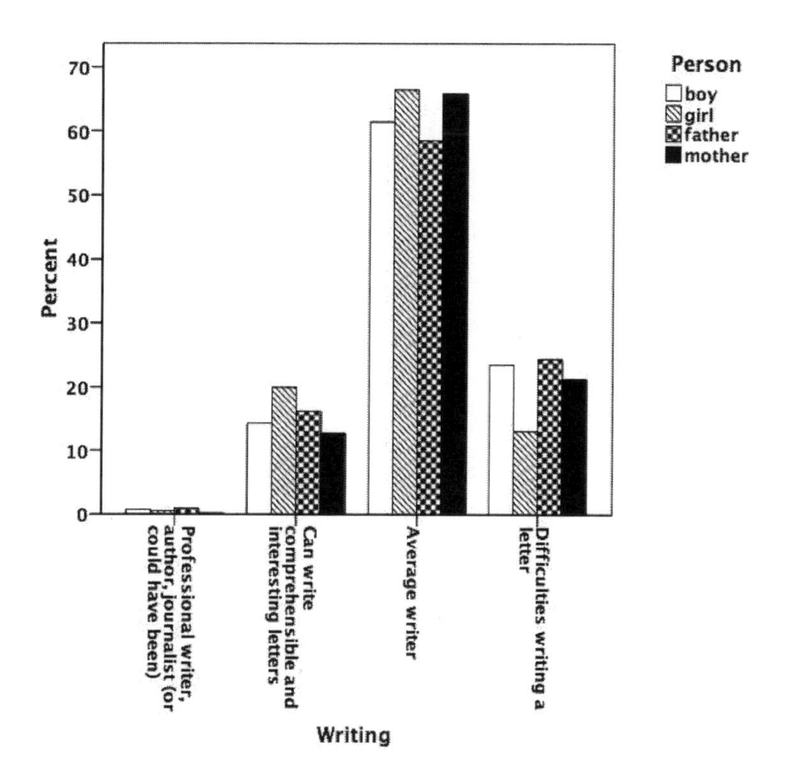

Figure 5.4
Prevalence of talent for writing.

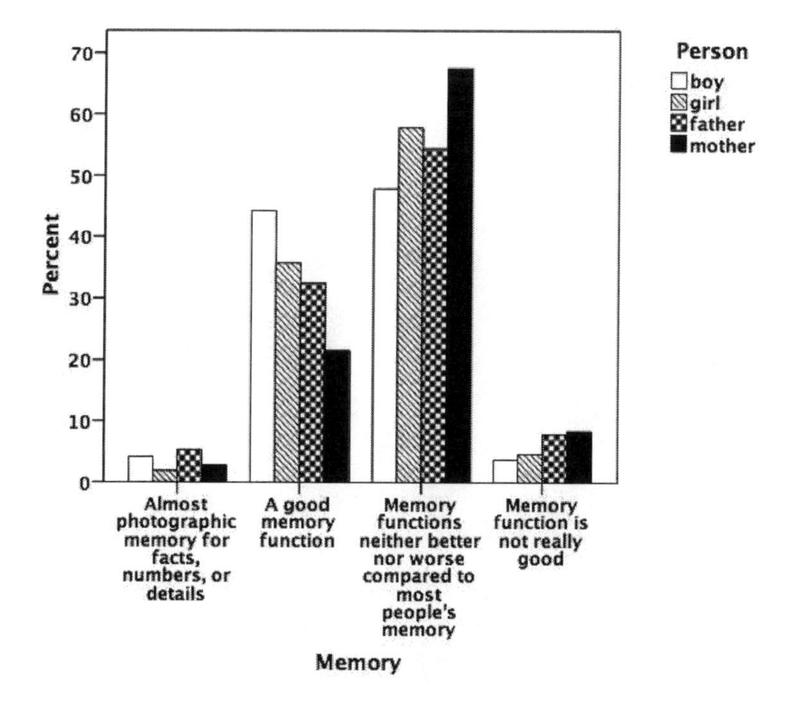

Figure 5.5
Prevalence of talent for memory.

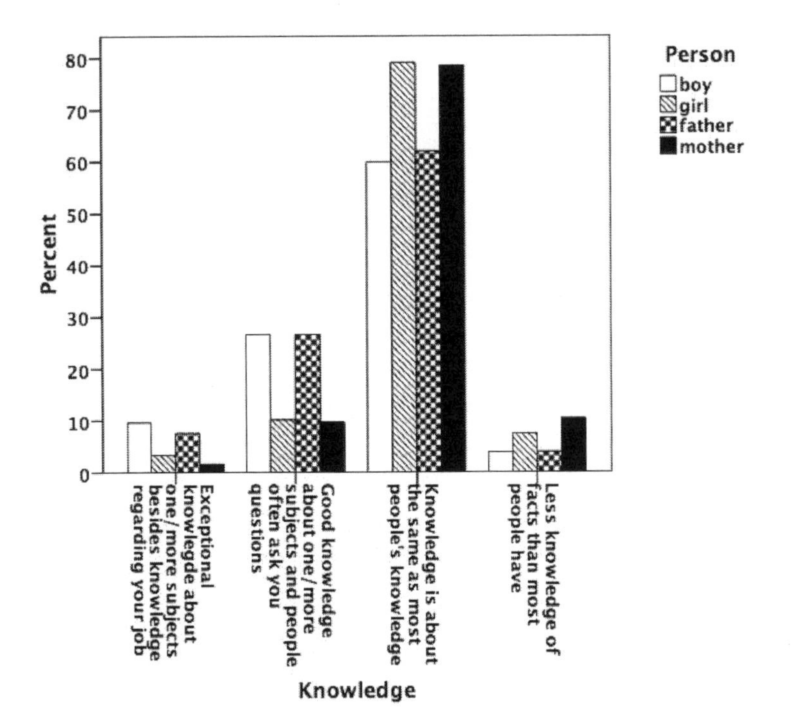

Figure 5.6
Prevalence of talent for knowledge.

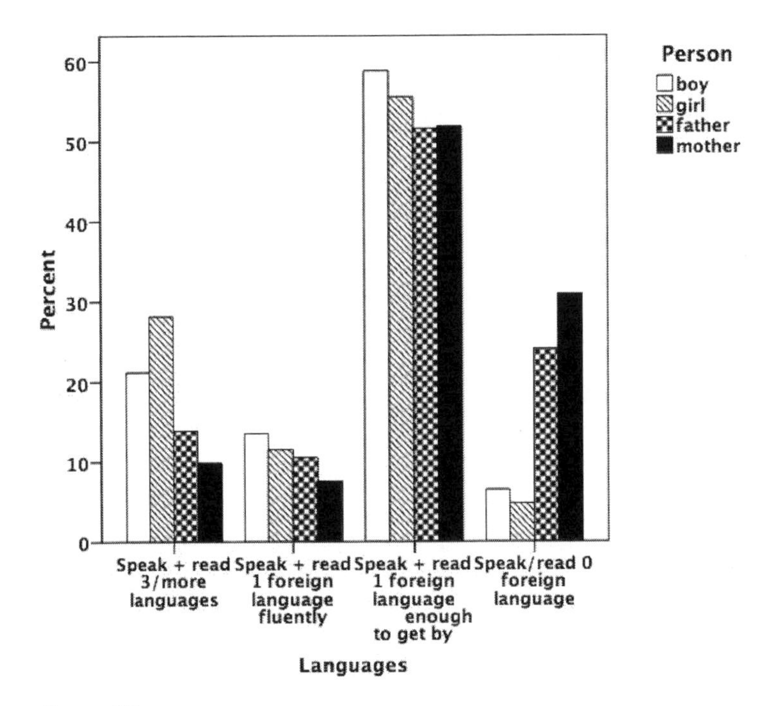

Figure 5.7
Prevalence of talent for languages.

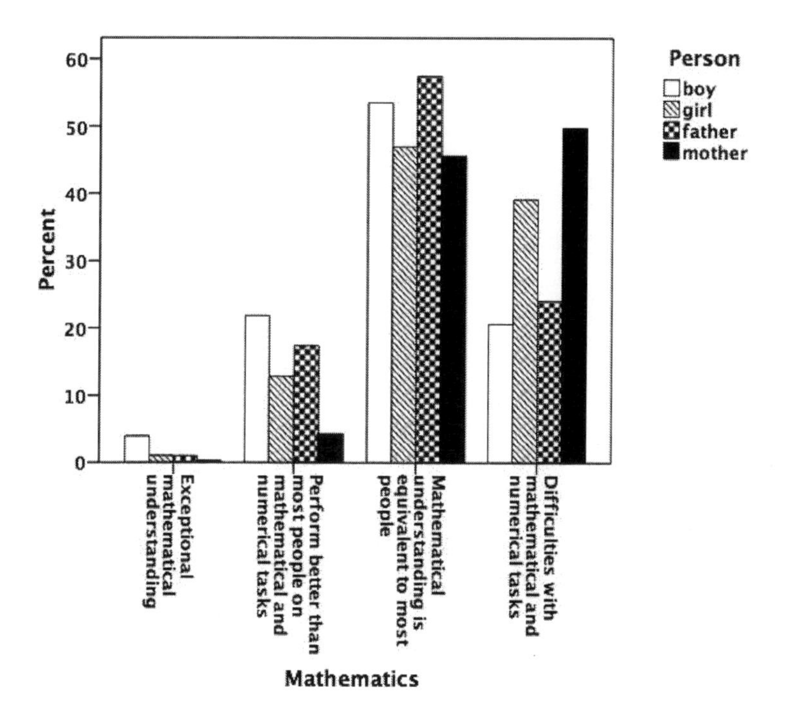

Figure 5.8
Prevalence of talent for mathematics.

in the offspring generation. From this figure it can also be seen that there are some clear generation and sex differences in prevalence of talent. Adolescent and young adult twins rate themselves more often as above average talented for Music, Arts, Memory, and Languages than their parents. We also see a clear tendency for men to report themselves more often as above average talented than women for Memory, Knowledge, Chess, and Mathematics. Adolescent girls rate themselves as most talented for Music, Arts, Writing, and Languages, compared to adolescent boys and their parents.

The polychoric correlations among family members are displayed in table 5.1. For all variables, MZ twin correlations exceeded the DZ twin correlations, suggesting genetic influences. MZ twin correlations ranged between 0.49 and 0.78. DZ twin correlations ranged from 0.09 up to 0.48. Only for Music was the DZ correlation larger than half the MZ correlation, which could be an indication of shared environmental influences. For some variables, including Chess, Writing, Memory, and Mathematics, the DZ correlation is smaller than half the MZ correlation, suggesting the influence of nonadditive genetic factors. When inspecting the parent-offspring correlations,

Table 5.1
Twin, parent-offspring, and spouse correlations for various talents.

	MZ	DZ	Parent-Offspring	Spouses
Music	.78 (.73–.82)	.48 (.41–.54)	.32 (.28–.35)	.15 (.07–.22)
Arts	.64 (.57–.70)	.27 (.19–.35)	.26 (.21–.30)	.18 (.09–.26)
Chess	.52 (.44–60)	.09 (.02–.17)	.11 (.07–.15)	.11 (.04–.17)
Writing	.49 (.40–.57)	.15 (.08–.23)	.13 (.09–.17)	.07 (–.01–.13)
Memory	.49 (.40–.57)	.07 (–.01–.16)	.09 (.05–.12)	–.11 (–.18–.04)
Knowledge	.62 (.53–.69)	.28 (.19–.36)	.14 (.09–.18)	.18 (.10–.26)
Languages	.71 (.64–.77)	.37 (.30–.44)	.30 (.25–.33)	.48 (.42–.53)
Mathematics	.70 (.64–.75)	.20 (.13–.28)	.22 (.18–.25)	.14 (.07–.21)

Notes: Correlation estimates with 95 percent confidence intervals.

it becomes clear that they are all significantly larger than zero and often very similar to the DZ twin correlations, suggesting that differences in genetic effects across generations are unlikely. Importantly, the fact that both DZ correlations and parent-offspring correlations are not larger than half the MZ correlations strongly suggests that cultural transmission is an unlikely explanation for the familial clustering of talent behavior. Spouse correlations are mostly significantly different from zero, except for Memory. This indicates that for all aptitudes except Memory there is significant nonrandom mating. Nonrandom mating is largest for Languages (0.48).

Conclusions

To summarize, the familial clustering for a variety of talents, already reported as early as in the end of the nineteenth century by pioneer Sir Francis Galton, is confirmed in our study. The unique features of the parents-twins design made it possible to demonstrate that this familial clustering is unlikely to be explained by cultural transmission from one generation to the next, but is rather due to genetic transmission. This holds for all the talents that were included in this study: Music, Arts, Chess, Writing, Memory, Knowledge, Languages, and Mathematics. In addition, nonrandom mating was observed for all talents except Memory. Some evidence for nonadditive genetic influences was found for Chess, Writing, Memory, and Mathematics, as indicated by the low DZ correlations, compared to the MZ correlations. Parent-offspring correlations for these talents were similarly low. This suggests that the familial correlations that were observed are best explained by genetic inheritance, that genetic effects include nonadditivity, and that environmental influences do not contribute to familial resemblance over and above the resemblance that is accounted for by genetic effects.

Our findings indicate that children who grow up with talented parents also are more often exposed to favorable environmental factors. There is no evidence for a "main effect" of cultural transmission, but this double advantage will often offer children with innate talent the opportunity to practice the skills that are required for the expression of talent. Moreover, those with talent will benefit from practice (for example, for music and chess) and to experience the rewards that practice can bring them. Unlike children without any potential for a particular talent, the reward they experience may act as a stimulus to continue practicing, whereby individual differences in practicing themselves become influenced by genotype.

Our findings also have implications for gene-finding studies. Since talented behavior is so clearly heritable, what are the genes that constitute this heritability? Recently, genome-wide association (GWA) studies have turned out to be quite successful for some complex polygenic physical and medical traits, such as height, body mass index, and type II diabetes (Lango Allen et al., 2010; Saxena et al., 2007; Speliotes et al., 2010; Yang et al., 2010). GWA studies have more difficulties in pinpointing the genetic variants that influence more complex cognitive, affective, and behavioral phenotypes, although some progress has been made (e.g., De Moor et al., 2010; Sullivan et al., 2009; Vink et al., 2009). For example, GWA studies for intelligence have come up with a limited number of genetic variants of small effect that may influence intelligence, but a recent study shows that the genome-wide polymorphisms are able to explain a large portion of the heritability of intelligence (Davies et al., 2011). To our knowledge, no such studies have been conducted for the talents discussed in this chapter. Yet, the findings concerning general intelligence could provide important information on the underlying molecular genetic basis of some of the talents presented here, such as Memory, Knowledge, Languages, and Mathematics.

References

Alarcon, M., Knopik, V. S., & DeFries, J. C. (2000). Covariation of mathematical achievement and general cognitive ability in twins. *Journal of School Psychology, 38,* 63–77.

Baharloo, S., Service, S. K., Risch, N., Gitschier, J., & Freimer, N. B. (2000). Familial aggregation of absolute pitch. *American Journal of Human Genetics, 67,* 755–758.

Benbow, C. P., & Lubinski, D. (1993). Psychological profiles of the mathematically talented: Some gender differences and evidence supporting their biological basis. In K. Ackerill (Ed.), *The origins and development of high ability* (pp. 44–66). New York: John Wiley.

Boomsma, D. I., Busjahn, A., & Peltronen, I. (2002). Classical twin studies and beyond. *Nature Reviews: Genetics, 3,* 872–882.

Boomsma, D. I., de Geus, E. J. C., Vink, J. M., Stubbe, J. M., Distel, M. A., Hottenga, J. J., et al. (2006). Netherlands Twin Register: From twins to twin families. *Twin Research and Human Genetics, 9,* 849–857.

Bouchard, T. I., Jr. (1998). Genetic and environmental influences on adult intelligence and special mental abilities. *Human Biology, 70,* 257–279.

Coon, H., & Carey, G. (1989). Genetic and environmental determinants of musical ability in twins. *Behavior Genetics, 19,* 183–193.

Dale, P. S., Harlaar, N., Haworth, C. M. A., & Plomin, R. (2010). Two by two: A twin study of second-language acquisition. *Psychological Science, 21,* 635–640.

Dale, P. S., Simonoff, E., Bishop, D. V. M., Eley, T. C., Oliver, B., Price, T. S., et al. (1998). Genetic influence on language delay in 2-year old children. *Nature Neuroscience, 1,* 324–328.

Davies, G., Tenesa, A., Payton, A., Yang, J., Harris, S. E., Liewald, D., et al. (2011). Genome-wide association studies establish that human intelligence is highly heritable and polygenic. *Molecular Psychiatry, 16,* 996–1005.

De Bruin, A. B., Smits, N., Rikers, R. M., & Schmidt, H. G. (2008). Deliberate practice predicts performance over time in adolescent chess players and drop-puts: A linear mixed models analysis. *British Journal of Psychology, 99,* 473–497.

De Moor, M. H. M., Costa, P. T., Terracciano, A., Krueger, R. F., de Geus, E. J. C., Tanaka, T., et al. (2010). Meta-analysis of genome-wide association studies for personality. *Molecular Psychiatry*; Epub ahead of print. doi: 10.1038/mp.2010.

Drayna, D., Manichaikul, A., de Lange, M., Snieder, H., & Spector, T. (2001). Genetic correlates of musical pitch recognition in humans. *Science, 291,* 1969–1972.

Eaves, L. J. (1972). Computer simulation of sample size and experimental design in human psychogenetics. *Psychological Bulletin, 77,* 144–152.

Eaves, L. J., Last, K., Martin, N. G., & Jinks, J. L. (1977). A progressive approach to non additivity and genotype-environmental covariance in the analysis of human differences. *British Journal of Mathematical and Statistical Psychology, 30,* 1–42.

Eaves, L. J., Fulker, D. W., & Heath, A. C. (1989). The effects of social homogamy and cultural inheritance on the covariances of twins and their parents: A LISREL model. *Behavior Genetics, 19,* 113–122.

Falconer, D. S., & Mackay, T. F. C. (1996). *An introduction to quantitative genetics*. Essex: Longmans, Green, Harlow.

Finkel, D., Pedersen, N., & McGue, M. (1995). Genetic influences on memory performance in adulthood: Comparison of Minnesota and Swedish twin data. *Psychology and Aging, 10*, 437–446.

Fulker, D. W. (1989). Genetic and cultural transmission in human behavior. In B. S. Weir, E. J. Eissen, M. M. Goodman, & G. Namkoong (Eds.), *Proceedings of the Second International Conference on Quantitative Genetics* (pp. 318–340). Sunderland, MA: Sinauer.

Fuller, J. L., & Thompson, W. R. (1978). *Foundations of behavior genetics*. Saint Louis: Mosby.

Galton, F. (1874). *English men of science: Their nature and nurture*. London: Macmillan.

Galton, F. (1869). *Hereditary genius*. London: Macmillan.

Heath, A. C., & Eaves, L. J. (1985). Resolving the effects of phenotype and social back-ground on mate selection. *Behavior Genetics, 15*, 15–30.

Howe, M. J. A., Davidson, J. W., & Sloboda, J. A. (1998). Innate talents, reality or myth. *Behavioral and Brain Sciences, 21*, 399–407.

Koopmans, J. R., van Doornen, L. J. P., & Boomsma, D. I. (1994). Smoking and sports participation. In U. Goldbourt, U. DeFaire, & K. Berg (Eds.), *Genetic factors in coronary heart disease* (pp. 217–235). Dordrecht: Kluwer Academic.

Lango Allen, H., Estrada, K., Lettre, G., Berndt, S. I., Weedon, M. N., Rivadeneira, R., et al. (2010). Hundreds of variants clustered in genomic loci and biological pathways affect human height. *Nature, 467*, 832–838.

Lubinsky, D., Benbow, C. P., Webb, R. M., & Bleske-Rechek, A. (2006). Tracking exceptional human capital over two decades. *Psychological Science, 17*, 194–199.

Maes, H. H., Neale, M. C., Medland, S. E., Keller, M. C., Martin, N. G., Heath, A. C., et al. (2009). Flexible Mx specification of various extend twin kinship designs. *Twin Research and Human Genetics, 12*, 26–34.

Martin, N. G., Boomsma, D. I., & Machin, G. A. (1997). A twin-pronged attack on complex traits. *Nature Genetics, 17*, 387–392.

McGue, M., Hirsch, B., & Lykken, D. T. (1993). Age and the self-perception of ability: A twin study analysis. *Psychology and Aging, 8*, 72–80.

Neale, M. C., Boker, S. H., Xie, G., & Maes, H.H. (2003). *Mx: Statistical modeling* (6th Ed.). Richmond: Virginia Commonwealth University.

Neale, M. C., & Cardon, L. R. (1992). *Methodology for genetic studies of twins and families*. Dordrecht: Kluwer Academic.

Newbury, D. F., Bishop, D. V. M., & Monaco, A. P. (2005). Genetic influences on language impairment and phonological short-term memory. *Trends in Cognitive Sciences, 9*, 528–534.

Plomin, R., DeFries, J. C., McClearn, G. E., & McGuffin, P. (2008). *Behavioral genetics* (5th Ed.). New York: Worth Publishers.

Polgár, L. (1989). *Nevelj zsenit! (Bring up genius!)*. Budapest: Kossuth Kiadó ZRT.

Reynolds, C. A., Barlow, T., & Pedersen, N. L. (2006). Alcohol, tobacco, and caffeine use: Spouse similarity processes. *Behavior Genetics, 36*, 201–215.

Rijsdijk, F. V., Vernon, P. A., & Boomsma, D. I. (2002). Application of hierarchical genetic models to Raven and WAIS subtests: A Dutch twin study. *Behavior Genetics, 32*, 199–210.

Ruthsatz, J., Detterman, D., Griscom, W., & Collins, J. M. (2008). Becoming an expert in the musical domain: It takes more than just practice. *Intelligence, 36*, 330–338.

Saxena, R., Voight, B. F., Lyssenko, V., Burtt, N. P., de Bakker, P. I. W., Chen, H., et al. (2007). Genome-wide association analysis identifies loci for type 2 diabetes and triglyceride levels. *Science, 316*, 1331–1336.

Simonton, D. K. (1999). *Origins of genius: Darwinian perspectives on creativity*. Oxford: Oxford University Press.

Speliotes, E. K., Willer, C. J., Berndt, S. I., Monda, K. L., Thorleifsson, G., Jackson, A. U., et al. (2010). Association analyses of 249,796 individuals reveal 18 new loci associated with body mass index. *Nature Genetics, 42*, 937–948.

Spinath, F. M., Price, T. S., Dale, P. S., & Plomin, R. (2004). The genetic and environmental origins of language disability and ability: A study of language at 2, 3, and 4 years of age in a large community sample of twins. *Child Development, 75*, 445–454.

Stromswold, K. (2001). The heritability of language: A review and meta-analysis of twin, adoption, and linkage studies. *Language, 77*, 647–723.

Sullivan, P. F., de Geus, E. J. C., Willemsen, G., James, M. R., Smit, J. H., Zandbelt, T., et al. (2009). Genomewide association for major depressive disorder: A possible role for the presynaptic protein piccolo. *Molecular Psychiatry, 14*, 359–375.

Thompson, L. A., Detterman, D. K., & Plomin, R. (1991). Associations between cognitive-abilities and scholastic achievement-genetic overlap but environmental differences. *Psychological Science, 2*, 158–165.

van Grootheest, D. S., van den Berg, S. M., Cath, D. C., Willemsen, G., & Boomsma, D. I. (2008). Marital resemblance for obsessive-compulsive, anxious, and depressive symptoms in a population-based sample. *Psychological Medicine, 38*, 1731–1740.

van Leeuwen, M., van den Berg, S. M., & Boomsma, D. I. (2008). A twin-family study of general IQ. *Learning and Individual Differences, 18*, 76–88.

Viding, E., Price, T. S., Spinath, F. M., Bishop, D. V. M., & Dale, P. S. (2003). Genetic and environmental mediation of the relationship between language and nonverbal impairment in 4-year-old twins. *Journal of Speech, Language, and Hearing Research: JSLHR, 46*, 1271–1282.

Vink, J. M., Smit, A. B., de Geus, E. J., Sullivan, P., Willemsen, G., Hottenga, J. J., et al. (2009). Genome-wide association study of smoking initiation and current smoking. *American Journal of Human Genetics, 84*, 367–379.

Vinkhuyzen, A. A. E., van der Sluis, S., Posthuma, D., & Boomsma, D. I. (2009). The heritability of aptitude and exceptional talent across different domains in adolescents and young adults. *Behavior Genetics, 9*, 380–392.

WAIS-III. (1997). Manual. Dutch version. Lisse: Swets and Zeitlinger.

Wijsman, E. M., Robinson, N. M., Ainsworth, K. H., Rosenthal, E. A., Holzman, T., & Raskind, W. H. (2004). Familial aggregation patterns in mathematical ability. *Behavior Genetics, 24*, 51–62.

Willemsen, G., Posthuma, D., & Boomsma, D. I. (2005). Environmental factors determine where the Dutch live: Results from the Netherlands Twin Register. *Twin Research and Human Genetics, 8*, 312–317.

Yang, J., Benyamin, B., McEvoy, B. P., Gordon, S., Henders, A. K., Nyholt, D. R., et al. (2010). Common SNPs explain a large proportion of the heritability for human height. *Nature Genetics, 42*, 565–569.

III Neuropsychology

6 Art and Dementia: How Degeneration of Some Brain Regions Can Lead to New Creative Impulses

Indre V. Viskontas and Bruce L. Miller

The study of individuals with dementia has largely been focused on describing the deficits and changes in behavior that accompany neural degeneration. Most studies have compared affected individuals with their healthy counterparts, and emphasize loss rather than change. Certainly, the loss of cognitive functions such as memory, language, and decision-making devastates patients and their families. But in some cases, patients and their families develop coping strategies that lead to the paradoxical emergence of new and positive behaviors such as gardening, painting, and other forms of personal expression. These cases provide insight not only into the capacity of the mind to adapt and survive, but also into the creative process and the drive to communicate. As artistic expression is deeply personal and subjective, the study of creativity in patients with dementia will need to focus on the individual as well as the group.

Defining and measuring creativity has proven to be a monumental task. Taken together with the considerable individual differences in response to brain damage, and the subjectivity of artistic expression, evidence unambiguously demonstrating increased creativity in patients with degenerative dementia is scarce. Here, we focus on the emergence of previously unrecognized artistic and musical creativity in the context of neurodegenerative illnesses such as frontotemporal dementia (FTD) (Miller et al., 1998, 2000). Understanding which skills are retained or even emerge in the setting of dementia not only aids in differential diagnosis, but can also help clinicians design appropriate treatment options and diminish the frustration that patients and their caregivers experience on a daily basis by finding activities that bring satisfaction and pleasure to patients. Furthermore, as age is the biggest risk factor in the development of dementia, highlighting the positive changes that can accompany progressive neural degeneration may also aid in decreasing the prevalence of prejudicial attitudes toward the elderly.

Neural Circuits Affected in Neurodegenerative Disease

Current directions in neuroscience research have begun to move away from regional specialization and toward an understanding of how cognitive and other functions are served by neural circuits. Neurodegenerative diseases such as Alzheimer's disease (AD) and FTD strike specific circuits; damage is not limited to one isolated region. Degeneration is not immediate, however, and the progressive nature of the disease provides neuroscientists with the opportunity to understand how circuits might cope with loss, and how the brain might reorganize functional specialization. In some patients, the reorganization can be striking, not only in terms of regions taking over certain functions, but also in the behavioral changes that reflect the mind's adaptation to functional loss. The study of neural plasticity is entering a golden age and our understanding of the neurobiology of creativity will certainly benefit from this new focus.

One of the recent advances in cognitive neuroscience has provided the tools with which scientists can observe connectivity between regions during different mental states. Fox and colleagues (2005) demonstrated that while certain brain regions such as the frontal and parietal cortices act in concert during tasks demanding focused attention, other regions show correlated activity only when the mind is "at rest," or not actively engaged in a task (Fox et al., 2005). The delineation of functional connectivity during different cognitive tasks is particularly fruitful in patients with neurodegenerative disease, as it allows scientists to understand the context in which new behaviors emerge and how functional reorganization with disease progression might occur.

Functional connectivity analyses in patients with neurodegenerative diseases have begun to demonstrate that these diseases target specific large-scale neural networks (Seeley et al., 2009). AD is characterized by episodic memory dysfunction, and results in atrophy in the "default mode network" (Fox et al., 2005), including the posterior cingulate/precuneus, medial temporal lobe, and lateral temporoparietal regions. FTD is an umbrella term for a group of diseases that result in progressive degeneration of the frontal and temporal lobes. These disorders are categorized into three subtypes: behavioral-variant frontotemporal dementia (bvFTD) causes primarily behavioral symptoms, including disinhibition, apathy, emotional blunting, lack of insight, and poor decision making, and anterior cingulate, frontoinsular, striatal, and frontopolar degeneration; semantic dementia (SD) patients show a loss of word and object meaning, or conceptual knowledge, with primarily left-temporal pole and subgenual cingulate atrophy;

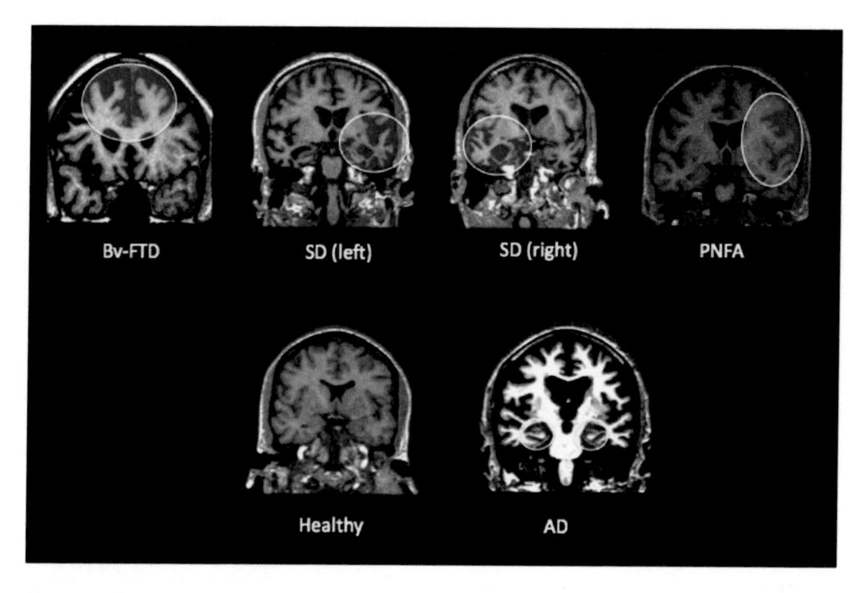

Figure 6.1
Coronal MRI images showing areas affected by neurodegenerative disease. BvFTD: behavioral-variant frontotemporal dementia; SD: semantic dementia, left: left hemisphere, right: right hemisphere; PNFA: progressive nonfluent aphasia; AD: Alzheimer's disease.

and progressive nonfluent aphasia (PNFA) leads to nonfluent, effortful, and agrammatic speech, caused by degeneration in the left frontal operculum, dorsal anterior insula, and precentral gyrus (see figure 6.1; for a review, see Boxer & Miller, 2005).

Networks Involved in Creativity

Although the relationship between these specific patterns of network degeneration and the emergence of creativity in specific patients remains murky, there is overlap between the networks affected by these diseases and those thought to underlie creativity. In particular, neuroscientists investigating creativity have moved away from the idea that creativity is primarily a right-hemisphere function (Bogen & Bogen, 1988), and toward the understanding that the interaction between the frontal and temporal lobes might be critical for generating and acting on creative impulses (Flaherty, 2005).

Providing a theoretical framework through which the neural networks involved in different types of creative output may be understood, Dietrich

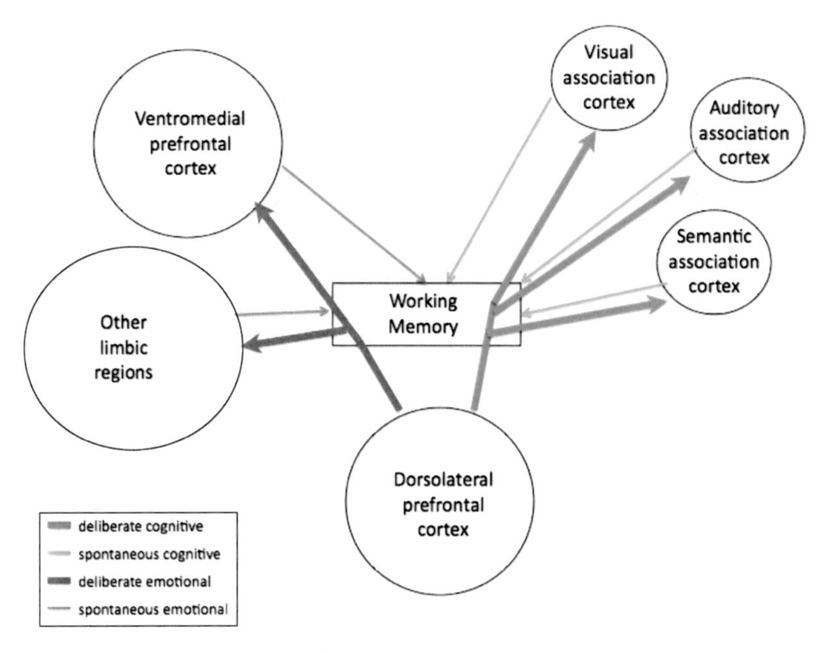

Figure 6.2
Types of creative output and the interactions between neural networks that support them. Thicker arrows indicate control, thinner arrows indicate information flow.

(2004) distinguishes between cognitive and emotional content, and spontaneous and deliberate creative processes. These two dimensions yield four distinct classes: insights can be cognitive and deliberate, cognitive and spontaneous, emotional and deliberate, or emotional and spontaneous. Furthermore, each class of creative thought is likely subserved by a different pattern of network interactions (see figure 6.2).

Cognitive deliberate insights likely rely heavily on dorsolateral prefrontal cortex (dlPFC) and its connections with domain-specific association cortex. Cognitive spontaneous insights reflect metabolic activity in the posterior association cortex. These thoughts seem to involve the basal ganglia, given its role in automatic behaviors, and enter consciousness, or working memory, during the periodic down-regulation of the frontal attentional system. Deliberate emotional insights require an interaction between the ventromedial PFC and the limbic system, including the amygdala and cingulate cortex. Spontaneous emotional insights are thought to be driven primarily by the activity of the limbic system, when the PFC is deactivated. Often, during biologically significant events and/or intense emotional experiences, information processed in emotion regions takes priority and overrides voluntary attention. Understanding how each of these types of insights is subserved by specific neural circuits will aid in the interpretation of the paradoxical facilitation of creativity with the degeneration of certain neural circuits.

The frontal cortex, in particular, receives information from regions throughout the brain, including motor, sensory, and emotion areas, and via reciprocal connections controls attention, language, memory, and other cognitive functions such as self-construct (Keenan et al., 2000), self-reflective consciousness (Courtney et al., 1998), complex social function (Damasio et al., 1994), cognitive flexibility, planning, willed action, source memory, and theory of mind. With changes in frontal neuroanatomy, one expects to see effects on creativity, although the exact nature of these effects is difficult to predict given the complexity of frontal lobe function.

Paradoxical Facilitation of Creativity with Neurodegenerative Disease

There have been some cases of patients in whom new or preserved musical or visual artistic abilities have been described in the setting of FTD (Miller et al., 1998, 2000). Given the central role that the frontal lobes play in creativity, the emergence of these abilities is often linked to degeneration of the anterior temporal lobe, particularly in the left hemisphere, as opposed to the frontal lobes. These patients, classified as having SD or PNFA, show progressive loss of conceptual knowledge or other language skills, and their creative output tends to be devoid of verbal or symbolic content (Miller et al., 2000).

These patients tend to paint realistic landscapes, animals, or detailed geometric designs that do not seem to contain abstract meaning beyond the superficial visual characteristics. Musicians continue to perform music, but when words are set to music, as is the case in popular songs, these patients do not use the words to express themselves. For example, one individual sang almost every day, and his primary pleasurable activity was leading elderly patients in a hospice setting in sing-alongs. He could sing any popular song in a large songbook, but when the words of a famous folk song ("My Bonny Lies Over the Ocean") were altered to the point of being incomprehensible, he did not miss a beat (Sacks, 2007). In our experience, however, much of the work created by these patients contains some recollections from the past and tends to avoid verbal or symbolic representations.

To explore the connection between anterior temporal lobe degeneration and visual art, Rankin and colleagues (2007) used standardized tests to probe creative cognition and novel tests of visual art creation in patients with dementia. Patients with both bvFTD and SD produced artwork that was rated as more bizarre and distorted than patients with AD and healthy age-matched controls. SD patients were more likely than participants from any of the other groups to produce drawings devoid of meaning, and to choose conventional or obvious markings, such as closing an open-ended figure in standardized visuospatial creativity tests, which involved creating drawings based on incomplete meaningless doodles. In contrast, the paintings that they created, and indeed much of the artwork from this patient group, was striking in the aesthetic dimension, characterized by the unorthodox use of vivid, unconventional colors and intricate, repetitive geometrical designs, underscoring the unique perception as well as the obsessive nature of their work. This pattern of findings—that SD patients are less creative on standardized visuospatial creativity tests but less conventional in their self-generated depictions of the world—underscores the difficulty in measuring and categorizing creativity. Great artists share their unique subjective experience of the world in a way that both resonates with and surprises their audience, and by that measure, some of these SD patient artists produced work of substantial artistic merit.

In line with their language deficits, however, SD patients showed decreased fluency, originality, and elaboration in concert with a tendency toward ending the test prematurely on verbal tests of creativity. These findings suggest that the enhancements in creativity that may accompany a particular degenerative pattern are likely to remain domain specific, at least in the case of artists with neurodegenerative disease. Just as mastery of skills in one domain such as music can improve creative output in that domain but not necessarily transfer to other artistic realms, so too might the paradoxical facilitation of creativity with neurological disease remain tied to a single type of output. Dean Simonton (1994) introduced the idea that creative works that meet both the uniqueness and appropriateness criteria are almost always a function of a highly developed skill used in a novel fashion. Before a novel creative masterpiece is produced, artists spend years developing the necessary skills in painting, music, writing, and science (Ericsson, 2008).

The development of these necessary skills is enabled by the plastic nature of the brain. Cortical plasticity facilitates fine-tuning of motor and sensory skills and allows the formation of novel associations that make up the building blocks of creativity, and has been observed in both the frontal and medial temporal lobes. For example, extensive practice of a musical instrument results in reorganization of motor and sensory regions involved in the activity, with a larger proportion of cortical real estate devoted to the sensory and motor processing of the relevant body part (Elbert & Rockstroh, 2004). The flexible and dynamic nature of brain regions such as the hippocampus and prefrontal cortex, where plasticity is most pronounced, enables the formation of new associations, which form the basis of creativity. When, as might be the case in patients with SD, these regions are disrupted in a progressive manner, the brain's efforts to compensate for loss of function might underlie the emergence of a new obsession and/or creative behavior.

The Case of Anne Adams

Recently, Seeley et al. (2008) described the work and disease progression of a visual artist, Anne Adams, who suffered from PNFA. Ms. Adams was a biologist by profession but painted as a hobby for much of her working life. As her disease progressed, her artistic drive increased so much that eventually painting became her primary daily activity. It is interesting to note that she might have chosen to spend her time engaged in any number of hobbies, but, like many similar patients, she chose to focus on creating art in the visual domain. Likewise, an SD patient who was a lawyer by profession and a musician by avocation chose to produce visual art when he became ill, rather than resorting to music, which would have been more familiar, or some other related activity (Miller et al., 1998).

Underscoring the paradoxical facilitation of creativity in the setting of disease, Anne's most stunning works of art were produced at a time when the disease had already made a significant impact on her brain function. In PNFA, the main site of atrophy is the left fronto-opercular cortex, leaving patients with effortful, nonfluent, and apractic speech (Gorno-Tempini et al., 2004), and difficulties with grammar and articulation. Adding another dimension to the story is the observation that one of her masterpieces was an elaborate painting inspired by Maurice Ravel's (1875–1937) famous *Bolero* (Seeley et al., 2008) composed at a time when Ravel himself is thought to have been suffering from a progressive aphasia. Ravel composed his most famous piece to the rhythms of the Moorish-Spanish dance, in the early stages of his disease, at age 53. The *Bolero* is a compulsively repetitive/perseverative piece (Amaducci et al., 2002), with the recurrence of a simple melodic theme, accompanied by a cyclical and simple bass line. Ravel considered the Bolero as "a piece for orchestra without music." "I'm going to try and repeat [the theme] a number of times without any development, gradually increasing the orchestra as best I can" (Orenstein, 1991).

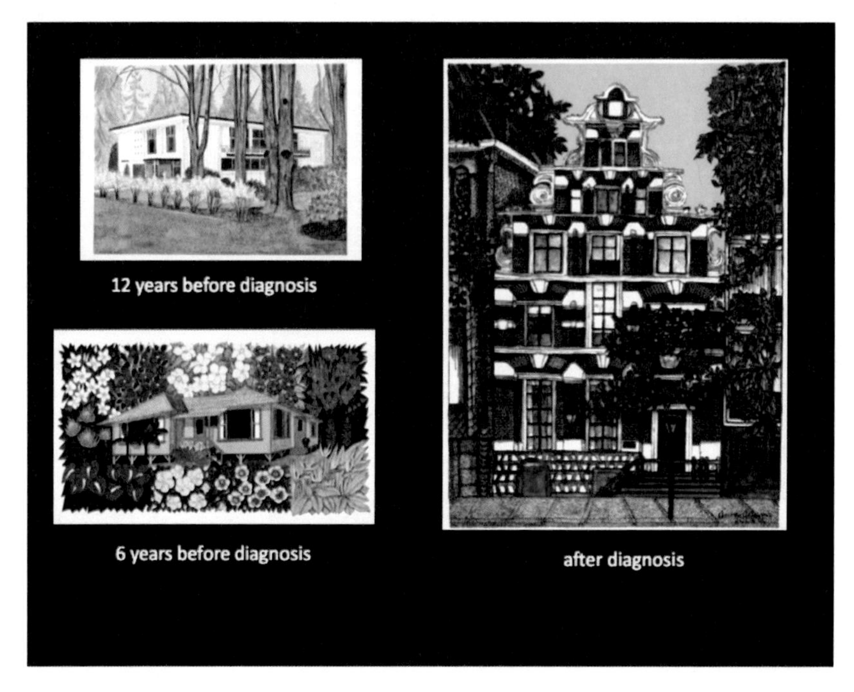

Figure 6.3
Representative artwork by Anne Adams as her disease progressed. Used with permission.

Anne Adams was inspired to paint the *Bolero*, although she was not aware that she and Ravel shared a similar illness. She painted the piece in a precise and compulsive fashion, representing each repetition of the melodic theme as a sequence of squares in a set of rows, with the height of each row signifying the increasing texture and volume of the orchestra. A retrospective of her art production throughout her lifetime shows a movement away from symbolic representations and toward detailed, obsessive interest in geometric patterns (see figure 6.3).

The case of Anne Adams is fascinating not only from the behavioral perspective, but also in terms of the structural and functional characteristics of her brain. Although her disease caused progressive degeneration primarily in the frontal cortex, Seeley and colleagues (2008) found a surprising surplus of gray matter volume and a functional enhancement in activation in heteromodal associative (intraparietal sulcus and superior parietal lobule: IPS/SPL) and polymodal (superior temporal sulcus: STS) neocortex in the right hemisphere. These areas have been shown to be involved in visuomotor search and attentional control (Corbetta & Shulman,

2002) and sensory transcoding, which are necessary for sight-reading music (Sergent et al., 1992). Professional musicians also have greater brain volume in these regions (Gaser & Schlaug, 2003).

This in-depth case study led to a larger study of visual attention in SD patients, with the goal of shedding some light on the perceptual experience of patients who might develop visual artistic creativity in the setting of neurodegenerative disease. An exploratory study of visual attention in SD patients unearthed striking differences between areas of focus of SD patients, healthy controls, and other patient groups during freeform viewing of complex images. SD patients tended to avoid areas that grabbed the attention of other patients and healthy controls, such as the vanishing point in images that emphasized perspective, and instead seemed to be interested in dense textures and complex color patterns. Some representative examples may be seen in figure 6.4. Comparing data from nine healthy controls, nine AD patients, nine FTD patients, and seven SD patients, we

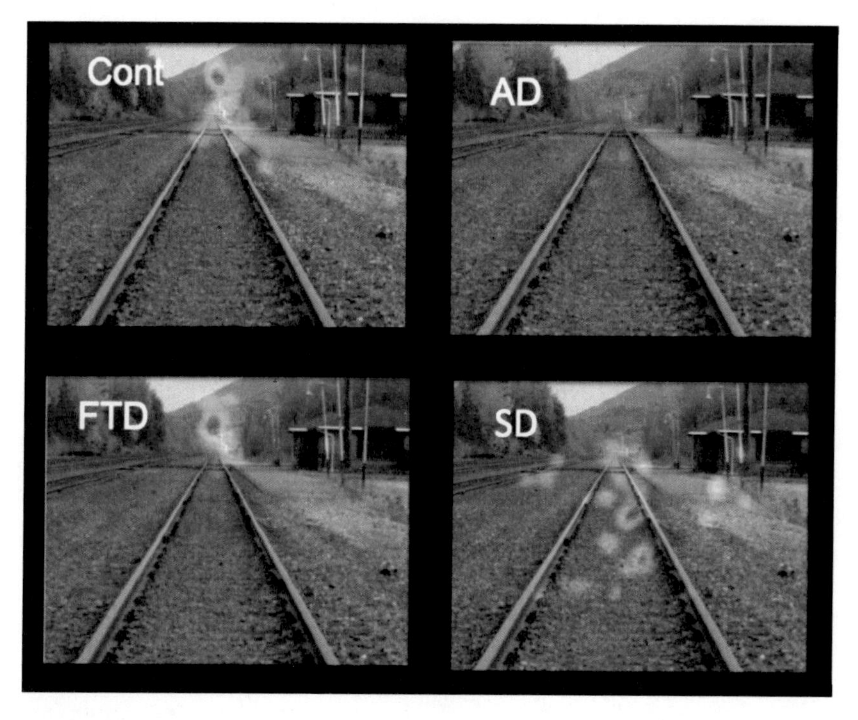

Figure 6.4
Mean eye movement fixation density plots from nine healthy controls (Cont), nine bvFTD patients (FTD), nine AD patients (AD), and seven SD patients (SD).

found no significant differences in terms of the mean number of fixation points, the mean length of the fixation time, or the distance between fixation points (length of saccades), but patients did differ from controls and each other in terms of where in the image they tended to spend the most time looking (I. V. Viskontas & B. L. Miller, unpublished raw data).

Intrigued by these findings, we designed a study to investigate how our patient groups might differ on a standard visual search task. To that end, we adapted a task pioneered by Anne Treisman and colleagues (1977) in which participants are asked to search an array of items for a target (Viskontas et al., 2011). When the target and distracters have distinguishable features, the search is conducted in parallel, leading to the experience of the target "popping out" of the array, and increasing the number of distracters has little or no effect on response time or accuracy. When, however, the target and distracters share features, the search is conducted by serially investigating each item, and therefore the response time in the task correlates positively with the number of items in the array, and participants are more likely to make errors as the number of distracters increases (Treisman & Gelade, 1980). In addition to recording response time and accuracy, we also monitored participants' eye movements and used voxel-based morphometry (VBM) to look for differences in brain volume that might account for differences in behavior.

Congruent with our observations that SD patients show preserved, if not enhanced, performance in tasks relying heavily on visual search such as completing jigsaw puzzles, playing solitaire, finding coins, and gardening, we found that SD patients were not only as accurate as healthy controls during both pop-out and serial search (both groups performed at ceiling), but they were significantly *faster* under the most difficult serial search condition. By contrast, AD patients showed significant impairments even under the easiest conditions and chance performance during serial search with an array size of thirty items. FTD patients were significantly worse than controls only in the most difficult serial search condition, and, like controls, their response times increased with more distracters in serial search. SD patients, in contrast, showed no increase in response time when the display size was doubled from fifteen to thirty items in the serial search condition, and found the target more quickly than all other groups even *while looking at a smaller proportion of the image.* That is, in the context of visual search tasks, SD patients may have a more efficient search strategy that allows them to ignore distracters and find the target more quickly than healthy people.

Performance on the serial search task correlated with gray matter volume in a dorsal frontoparietal network (including a region in the superior parietal lobe, the precuneus, the middle frontal gyrus, and higher visual regions in the occipital lobe) that is commonly associated with visual attention and which is spared in the setting of SD. The serial search task relies heavily on the binding of visual features into a coherent trace and the region of the parietal lobe that correlated with better performance in this task has also been shown to result in binding problems when it is damaged, as in Balint's syndrome or hemispatial neglect (Chan et al., 2003). Furthermore, hyperactive binding, as seen in individuals with synesthesia, has also been correlated with neural activity and more gray matter in the parietal lobe (Weiss & Fink, 2009). There was no appreciable difference in gray matter in this region between SD patients and healthy controls, supporting the idea that the enhancement in visual search function seen in SD patients results from increased functional connectivity rather than structural differences. This hypothesis has recently gained more support as enhancements in dorsal parietal network connectivity have been reported in patients with FTD (Zhou et al., 2010). Taken together, the results of this comprehensive study of visual search in SD patients suggests that the paradoxical enhancement of visual artistry seen in certain patients might reflect a shift in the attentional space from language to vision, resulting from an enhancement in connectivity in the dorsal parietal network.

Motivation and Creativity

Despite their progressive neurodegeneration, SD and PNFA patients who showed enhanced creativity with disease progression share many features with historical figures with great creativity. For example, they show more interest and/or obsessions with their subjects, they neglect social and occupational responsibilities in favor of their art production, and they continue to produce work even in the absence of any encouragement or support from others (Miller et al., 1998). This observation leads to questions regarding the underlying motivating factors in these patients, and artists in general, to spend time creating art. Notably, patients with emergent new or preserved old artistic skills generally do not show extensive frontal lobe degeneration (Miller et al., 2000). This observation is in line with the wealth of research on creativity in neurologically intact individuals, which suggests that certain components of frontal lobe function are critical for the many organizational and motivational components of creativity (Chávez-Eakle et al., 2007).

The involvement of the PFC in creativity underscores the important role that motivation plays in generating and sustaining creative behavior. While creativity is described as a virtue in a wide variety of careers, the creative life is generally not financially rewarding. In fact, a common view among social psychologists is that monetizing creative output can destroy or disrupt the very process that leads to innovation (Amabile et al., 1986). The relationship between reward and creativity remains murky, as studies have found both negative and positive relationships between incentives and output (Eisenberger & Cameron, 1996). What's clear, however, is that some tasks seem to be rewarding in and of themselves; as Harlow et al. (1950) first demonstrated, even monkeys will repeatedly solve mechanical puzzles even when the solution does not lead to any food or other extrinsic reward. Some activities are simply intrinsically rewarding, and many of these activities seem to have some component of creativity such as spontaneity, making new associations, or using one's imagination. A meta-analysis conducted by Cameron and Pierce (1994) suggests that while verbal rewards increase the time that individuals spend on intrinsically rewarding tasks, tangible rewards such as money, gold stars, or awards tend to decrease the time spent on task once those rewards are removed. Furthermore, this effect is seen only when tangible rewards are expected, and disappears when the rewards are spontaneously given after the task is accomplished. One of the most widely accepted explanations for these effects is that adding tangible extrinsic rewards to the equation shifts the focus from the task itself and onto the narrow scope of the goal (Amabile et al., 1986).

Flow

Mihaly Csikszentmihalyi refers to the pleasurable state that motivates the engagement of these activities as "flow" (Csikszentmihalyi, 1991), and suggests that nine separable coexisting factors are necessary to create it: (1) the delineation of clear goals, wherein expectations and rules are discernible, goals are attainable, and goals align appropriately with one's skill set and abilities; (2) a high degree of concentration on a limited number of items; (3) a loss of the feeling of self-consciousness, via the merging of action and awareness; (4) a distorted sense of subjective time; (5) direct and immediate feedback, which makes successes and failures apparent, so that behavior can be adjusted as needed; (6) a balance between ability level and challenge, such that the activity is neither too easy nor too difficult; (7) a sense of personal control over the situation or activity; (8) the activity is intrinsically rewarding, so that action feels effortless; and finally, (9) the activity is all-encompassing.

The dominance of the visual domain and the perseverative or obsessive behaviors characteristic of patients with SD can also be seen in the large proportion (~25 percent) of these patients for whom the completion of jigsaw puzzles becomes an important, or even the primary activity of daily living (Green & Patterson, 2009), likely because they are able to achieve "flow" by working on these puzzles. In a controlled study of jigsaw puzzle activity, Green and Patterson (2009) found that SD patients have preserved jigsaw puzzle completion skills, sometimes even above and beyond the performance of age-matched controls. In particular, these patients showed better performance than their healthy counterparts in completing "reality-disrupted" puzzles, in which expectations based on knowledge of the real world can interfere with puzzle completion, and on "grain" jigsaw puzzles, where conceptual knowledge does not benefit performance. Most encouraging, from a clinical standpoint, SD patients, who often exhibit flat affect and demeanor, have been observed to display signs of pleasure and pride during the completion of jigsaw puzzles, even if the activity was not a primary one for that particular person, suggesting that this task might be a good candidate for enabling the pleasurable state of flow which might drive the quest to create in some patients with SD.

The Role of Inhibition in Creativity

Besides having the necessary skills in a given field, creative individuals often lack inhibition in both behavior and cognition (Martindale & Hines, 1975; Martindale, 1999), a trait seen in many patients who show artistic expression in the setting of neurodegenerative disease. Anecdotally, creative people describe themselves as lacking self-control, and the creative process is sometimes described as effortless and without deliberation (Csikszentmihalyi, 1996). In line with the network view of neurodegenerative disease and higher cognitive function, Miller and colleagues (2000) have suggested that the mechanism by which creativity is unleashed in the setting of left anterior temporal lobe degeneration might involve releasing right dorsal parietal network from inhibition. The psychiatric disorders that have been also associated with creativity, such as bipolar disorder, usually include impulsive behaviors in their symptomatology (Martindale, 1971). It follows, then, that decreased activation in the frontal lobes, particularly on the left side, might be the source of disinhibition that has been associated with creativity. Support for this idea can be found in a doctoral dissertation by Hudspeth (1985), who found that more creative people show higher amplitude frontal lobe theta wave activity, which presumably indicates lower frontal lobe function. These results are consistent with findings reported by Carlsson et al. (2000), who found that decreases in regional blood flow (rCBF) in both the left and right superior frontal lobes correlated with superior performance on the Alternate Uses Task. Taken together, these findings highlight the complexity of the creative process and demonstrate that whereas the frontal lobes are involved in creative cognition and the organization of deliberate creative behaviors as described in the model presented earlier in this chapter, many of the spontaneous aspects of creative output seem to be supported by regions outside of the frontal cortex. One might even go a step further and suggest that under some conditions the frontal lobes need to be "turned off" in order to facilitate the emergence of these spontaneous behaviors, or that these behaviors emerge during the periodic down-regulation of frontal lobe function in healthy individuals (Dietrich, 2004).

The hypothesis that deactivation of frontal cortex might enhance creativity was tested by Snyder and colleagues (2003), who used repetitive transcranial magnetic stimulation (rTMS), a noninvasive method of transiently deactivating parts of the cortex in awake humans. Indeed, rTMS of the left frontotemporal lobe enhanced the creative components of certain skills, such as drawing (though not necessarily drawing ability per se, but rather the ability to capture perspective, kinetics, and certain highlighted details) and the ability to detect commonly overlooked duplicate words while proofreading. These proofreading skills rely on similar perceptual and attentional processes that are enhanced in SD patients. Caution must be used to interpret these results, however, as the effects were seen only in a subset of participants (4/11 in drawing, 2/11 in proofreading). Nevertheless, the results are consistent with the general notion that deliberate language-based focus can inhibit spontaneous creative thinking and that the absence of these constraints may enable the dynamic processes which lead to the novel recombination of ideas (Bristol & Viskontas, 2006).

Summary

Progressive neurodegenerative disease that disrupts the interactions between the frontal and the temporal, parietal, and occipital lobes, or between the dominant and nondominant hemispheres, has been shown to affect creativity in a myriad of ways. Diminished language function via neurodegenerative diseases that target the left frontal or left anterior temporal lobes sometimes leads to the emergence of previously unrecognized visual and musical creativity, possibly by facilitating function in posterior brain regions. Down-regulation of frontal function may enable spontaneous creative insights, but as patients with frontal lobe dysfunction demonstrate, the frontal lobes are necessary for many components of creativity, including organization, monitoring, and other executive functions. The importance of studying the paradoxical facilitation of behaviors that can help patients achieve "flow" is underscored by the observation that patients who engage in creative activities display many signs of improved quality of life. It is important to note, however, that while the art created by patients with neurodegenerative disease might be hailed as more creative by the artistic community, the brain networks underlying the drive to make these choices may not necessarily be the same as those upon which healthy artists rely.

References

Amabile, T. M., Hennessey, B. A., & Grossman, B. S. (1986). Social influences on creativity: The effects of contracted-for reward. *Journal of Personality and Social Psychology, 50,* 14–23.

Amaducci, L., Grassi, E., & Boller, F. (2002). Maurice Ravel and right-hemisphere musical creativity: Influence of disease on his last musical works? *European Journal of Neurology, 9,* 75–82.

Bogen, J. E., & Bogen, G. M. (1988). Creativity and the corpus callosum. *Psychiatric Clinics of North America, 11,* 293–301.

Boxer, A. L., & Miller, B. L. (2005). Clinical features of frontotemporal dementia. *Alzheimer Disease and Associated Disorders, 19,* S3–S6.

Bristol, A. S., & Viskontas, I. V. (2006). Dynamic processes within associative memory stores: Piecing together the neural basis of creative cognition. In J. C. Kaufman & J. Baer (Eds.), *Creativity, knowledge, and reason* (pp. 60–80). Cambridge: Cambridge University Press.

Cameron, J., & Pierce, W. D. (1994). Reinforcement, reward, and intrinsic motivation: A meta-analysis. *Review of Educational Research, 20,* 51–61.

Carlsson, I., Wendt, P. E., & Risberg, J. (2000). On the neurobiology of creativity: Differences in frontal activity between high and low creative subjects. *Neuropsychologia, 38,* 873–885.

Chan, R. C., Robertson, I. H., & Crawford, J. R. (2003). An application of individual subtest scores calculation in the Cantonese version of the Test of Everyday Attention. *Psychological Reports, 93,* 1275–1282.

Chávez-Eakle, R. A., Graff-Guerrero, A., García-Reyna, J. C., Vaugier, V., & Cruz-Fuentes, C. (2007). Cerebral blood flow associated with creative performance: A comparative study. *NeuroImage, 38,* 519–528.

Corbetta, M., & Shulman, G. L. (2002). Control of goal-directed and stimulus-driven attention in the brain. *Nature Reviews: Neuroscience, 3,* 201–215.

Courtney, S.M., Petit, L., Haxby, J.V., & Ungerleider, L.G. (1998). The role of prefrontal cortex in working memory: Examining the contents of consciousness. *Philosophical Transactions of the Royal Society, Series B: Biological Sciences, 353,* 1819–1828.

Csikszentmihalyi, M. (1991). *Flow: The psychology of optimal experience.* New York: Harper & Row.

Csikszentmihalyi, M. (1996). *Creativity.* New York: Harper Collins.

Damasio, H., Grabowski, T., Frank, R., Galaburda, A. M., & Damasio, A. R. (1994). The return of Phineas Gage: Clues about the brain from the skull of a famous patient. *Science, 264,* 1102–1105.

Dietrich, A. (2004). The cognitive neuroscience of creativity. *Psychonomic Bulletin & Review, 11,* 1011–1026.

Eisenberger, R., & Cameron, J. (1996). Detrimental effects of reward: Reality or myth? *American Psychologist, 51,* 1153–1166.

Ericsson, K. A. (2008). Deliberate practice and acquisition of expert performance: A general overview. *Academic Emergency Medicine, 15,* 988–994.

Elbert, T., & Rockstroh, B. (2004). Reorganization of human cerebral cortex: The range of changes following use and injury. *Neuroscientist, 10,* 129–141.

Flaherty, A. W. (2005). Frontotemporal and dopaminergic control of idea generation and creative drive. *Journal of Comparative Neurology, 493,* 147–153.

Fox, M. D., Snyder, A. Z., Vincent, J. L., Corbetta, M., Van Essen, D. C., & Raichle, M. E. (2005). The human brain is intrinsically organized into dynamic, anticorrelated functional networks. *Proceedings of the National Academy of Sciences of the United States of America, 102,* 9673–9678.

Gaser, C., & Schlaug, G. (2003). Brain structures differ between musicians and non-musicians. *Journal of Neuroscience, 23,* 9240–9245.

Gorno-Tempini, M. L., Dronkers, N. F., Rankin, K. P., Ogar, J. M., Phengrasamy, L., Rosen, H. J., et al. (2004). Cognition and anatomy in three variants of primary progressive aphasia. *Annals of Neurology, 55,* 335–346.

Green, H. A., & Patterson, K. (2009). Jigsaws: A preserved ability in semantic dementia. *Neuropsychologia, 47,* 569–576.

Harlow, H. F., Harlow, M. K., & Meyer, D. R. (1950). Learning motivated by a manipulation drive. *Journal of Experimental Psychology, 40,* 228–234.

Hudspeth, S. (1985). *The neurological correlates of creative thought.* Unpublished PhD dissertation. Los Angeles, CA: University of Southern California.

Keenan, J. P., Wheeler, M. A., Gallup, G. G., Jr., & Pascual-Leone, A. (2000). Self-recognition and the right prefrontal cortex. *Trends in Cognitive Sciences, 4,* 338–344.

Martindale, C. (1971). Degeneration, disinhibition, and genius. *Journal of the History of the Behavioral Sciences, 7,* 177–182.

Martindale, C. (1999). Biological bases of creativity. In R. J. Sternberg (Ed.), *Handbook of creativity* (pp. 137–152). Cambridge: Cambridge University Press.

Martindale, C., & Hines, D. (1975). Creativity and cortical activation during creative intellectual and EEG feedback tasks. *Biological Psychology, 3,* 91–100.

Miller, B. L., Boone, K., Cummings, J. L., Read, S. L., & Mishkin, F. (2000). Functional correlates of musical and visual ability in frontotemporal dementia. *British Journal of Psychiatry, 176,* 458–463.

Miller, B. L., Cummings, J., Mishkin, F., Boone, K., Prince, F., Ponton, M., et al. (1998). Emergence of artistic talent in frontotemporal dementia. *Neurology, 51,* 978–982.

Orenstein, A. (1991). *The ballets of Maurice Ravel: Creation and interpretation.* Burlington: Ashgate.

Rankin, K. P., Liu, A. A., Howard, S., Slama, H., Hou, C. E., Shuster, K., et al. (2007). A case-controlled study of altered visual art production in Alzheimer's and FTLD. *Cognitive and Behavioral Neurology, 20,* 48–61.

Sacks, O. W. (2007). *Musicophilia: Tales of music and the brain.* New York: Alfred A. Knopf.

Seeley, W. W., Matthews, B. R., Crawford, R. K., Gorno-Tempini, M. L., Foti, D., Mackenzie, I. R., et al. (2008). Unravelling *Bolero*: Progressive aphasia, transmodal creativity, and the right posterior neocortex. *Brain, 131,* 39–49.

Seeley, W. W., Crawford, R. K., Zhou, J., Miller, B. L., & Greicius, M. D. (2009). Neurodegenerative diseases target large-scale human brain networks. *Neuron, 62,* 42–52.

Sergent, J., Zuck, E., Terriah, S., & MacDonald, B. (1992). Distributed neural network underlying musical sight-reading and keyboard performance. *Science, 257,* 106–109.

Simonton, D. K. (1994). *Greatness: Who makes history and why.* New York: Guilford Press.

Snyder, A. W., Mulcahy, E., Taylor, J. L., Mitchell, D. J., Sachdev, P., & Gandevia, S. C. (2003). Savant-like skills exposed in normal people by suppressing the left fronto-temporal lobe. *Journal of Integrative Neuroscience, 2,* 149–158.

Treisman, A., Sykes, M., & Gelade, G. (1977). Selective attention and stimulus integration. In S. Dornic (Ed.), *Attention and performance VI.* Hillsdale, NJ: Erlbaum.

Treisman, A. M., & Gelade, G. (1980). A feature integration theory of attention. *Cognitive Psychology, 12,* 97–136.

Viskontas, I. V., Boxer, A. L., Fesenko, J., Matlin, A., Heuer, H. W., Mirsky, J., et al. (2011). Visual search patterns in semantic dementia show paradoxical facilitation of binding processes. *Neuropsychologia, 49,* 468–478.

Weiss, P. H., & Fink, G. R. (2009). Grapheme-colour synaesthetes show increased grey matter volumes of parietal and fusiform cortex. *Brain, 132,* 65–70.

Zhou, J., Greicius, M. D., Gennatas, E. D., Growdon, M. E., Jang, J. Y., Rabinovici, G. D., et al. (2010). Divergent network connectivity changes in behavioural variant frontotemporal dementia and Alzheimer's disease. *Brain, 133,* 1352–1367.

7 Biological and Neuronal Underpinnings of Creativity in the Arts

Dahlia W. Zaidel

General creative processes apply not only to the arts but also to science, technology, business, education, humor, interpersonal relationships, and many other domains of human expressions. The concept of creativity typically refers to the innovation of something new and positive for society, something that transcends the traditional and "received" knowledge. Even when the moment of innovation seems at times to be nonlinear, accidental, or to "come from nowhere," it comes on top of a body of mentally stored knowledge in the brain. Indeed, the backdrop for the creative innovation is the societal culture of the creating individual. Creativity also implies cognitive flexibility and rich associations among units of stored knowledge. How the cognitive departure from the norm is achieved, why some individuals in society can reach it, and what the underlying neuronal connectivity in the brain might be have long been sources of great interest and discussion.

Art creativity might be a special case of general creativity because with art, artistic talent and skill are critically interwoven into the artistic formula. Talent appears to have innate, inborn features, and skills can both be learned by the untalented as well as practiced by the talented. Where in the process of producing art does creativity come in: this is the principal question. Part of the answer might lie in the fact that spontaneous art production is an activity unique to humans and thus subject to its own distinctive rules of operation. Thus far, the evidence gathered from professional artists with brain damage does not point to any specific neuronal circuitry, hemispheric laterality, anatomical localization, or functional pathway that gives rise to creativity or to talent (Zaidel, 2005, 2010). Subsequent sections of this chapter discuss these findings and trace the biological and neuroanatomical backgrounds of creativity.

Biology: Comparative Considerations

Viewed within a biological perspective, the general notion of creativity, as in "innovation," is argued by some to not be unique to humans; if we look for the antecedents of general creativity, we find several examples of innovation by animals (Laland & Reader, 2010; Kaufman, Butt, Kaufman, & Colbert-White, 2011; Benson-Amram & Holekamp, 2012; Taylor, Miller, & Gray, 2012). As Bonner (1980) describes, there are several, by now classic, observations of innovation in animals. Hinde and Fisher (1951) described how titmice birds in the United Kingdom cleverly managed to peck holes in the aluminum foil lids of milk bottles left by the milk delivery person and then managed to lap up the milk; the practice began in one location but then spread to other locations in Britain. Similarly, researchers in Japan observed the behavior of a female monkey (named Imo) spontaneously washing sand off of her sweet potato in the river water before placing it in her mouth, and then this behavior was taken over by the rest of the monkey group (Kawai, 1965; Kawamura, 1959). Subsequently, the same female monkey innovated a method for washing sand off of wheat grains by dumping them in the river water and then scooping them from the surface (by which time they were all clean). Thus, human creativity, whether in the arts or not, has components that originated in biological ancestry.

Moreover, brain size strongly correlates with innovation in some animals; in birds, those regions are known as the hyperstriatum and neostriatum, whereas in primates the areas involve the isocortex and the striatum (Lefebvre, Reader, & Sol, 2004). These human brain regions are the association cortical areas. Looking at birds, a number of meta-analytic studies have found that deviations from typical behavior that enhance survival are associated with larger brains (Lefebvre et al., 2004). The rate of innovation is also highly related to tool use, learning, and, among birds, abilities dealing with seasonal changes. Indeed, according to one view, brain size evolution in birds has been driven by regions controlling behavior rather than by environmental changes (Wyles, Kunkel, & Wilson, 1983). By now, the countless observations of bird behavior, particularly with regard to innovation, strongly support this assumption (Reader & Laland, 2002; Laland & Reader, 2010).

With regard to primates, field observations have documented numerous instances of innovative behaviors (Byrne & Whiten, 1990; Goodall, 1986), typically in the context of deception rather than in technological skills. The driving forces behind the behaviors are hypothesized to be social. One of the hallmarks of primate behavioral evolution is the development of social interactions, interdependence, and complexity of hierarchy. Thus, survival depends heavily on cunning and flexibility of cognitive responses (Byrne, 2003; Byrne & Bates, 2010).

Intelligence Level and Creativity

Intelligence level plays a pivotal role in creativity. With animals, this is difficult to measure, but in humans we can measure it. Studying the history of a number of highly creative individuals in the arts and sciences, Howard Gardner (1994) unraveled a pattern of several shared factors. Some of those characteristics include at the very minimum a moderate level of intelligence. Robert Sternberg (1997; Sternberg & O'Hara, 1999) also included intelligence level in creativity and considered it to be a critical component; he listed motivation, knowledge, personality, cognition, and the environment as being important as well. The implication is that the creative process in individuals with compromised intelligence is seriously constricted.

Intelligence, as measured by the Intelligence Quotient (IQ), has been studied in recent years with an eye toward its neuroanatomical and neurofunctional underpinnings. Research with functional magnetic resonance imaging (fMRI) has revealed that the brain's functional and structural organization is different for those with high versus low intelligence in normal individuals (Fink & Benedek, this vol.; Jung & Haier, this vol.; Neubauer & Fink, 2009; Neubauer, Grabner, Fink, & Neuper, 2005). Genetic factors linking genes to intelligence have also recently been identified (Hulshoff Pol et al., 2006; Li et al., 2009), and this reinforces the notion of its neuroanatomical, neurophysiological, and neurofunctional underpinnings. At the same time, the findings of Deary and colleagues caution that genetics alone cannot explain expressions of intelligence (Deary, Penke, & Johnson, 2010). Cultural domains within which intelligence is expressed interact with the genetic bases (see Laland, Odling-Smee, & Myles, 2010; Laland, 2011). Indeed, creativity involves the mental possession of the cultural knowledge, be it specialized as scientific opinions about one issue or another, school movements in art, or financial business practices.

Although the genetics of intelligence continues to be researched and measured, the same cannot be said for the genetics of creativity. This is largely because quantifying creativity in ways that everyone can agree upon continues to be debated, whereas with intelligence we have actual numerical values for IQ, even while there is still debate over the meaning of the IQ itself. In any case, psychometric measures of creativity are predefined and reflect laboratory testing of creativity rather than spontaneous moments of "eureka" and insights experienced in daily life (whether by an artist, a scientist, or anyone else).

An important study, unrelated to art, by Jung and associates (Jung et al., 2009) attempted to link neuroanatomical regions with behavioral (psychometric) measures of creativity through the use of MRI. They found that volumetric cortical thickness in certain regions correlated with the psychometric measures of creativity. The thickness of the left lateral orbito-frontal region, the right angular gyrus, and the right cingulate cortex were highly correlated with creativity scores. A further recent study implicated the integrity of white matter in creativity as measured by psychometric tests (Jung, Grazioplene, Caprihan, Chavez, & Haier, 2010). Another creativity-related study (Fink et al., 2009) where subjects provided verbal alternate uses has shown increased activity in the left inferior parietal gyrus and angular gyrus. Viewed together, the neuroanatomical distance of these regions from each other and their laterality strongly suggest that neural connectivity is a critical component of the creative process. Whether or not all creative, original, innovative, groundbreaking activities are connectivity dependent versus region dependent remains to be explored in future studies.

Talent and Neuroanatomy

As stated in the opening paragraphs of this chapter, the issue of creativity in art is entangled with artistic talent and skill. What is artistic talent? Can it be separated from creativity? To what extent can talent be manipulated, modified, and made to grow as a function of creativity? At the very minimum, talent is an inborn ability to depict ideas in a representational way (on canvas or through any other medium), whether emanating from reality or from one's mind, in such a way that audiences are attracted to the representation. Obviously, artistic talent ranges from amateur to professional, from the occasional dabbler to the prolific artisan. The creative process can apply at each level of the talent continuum with the impact of the creative accomplishment likely to be incrementally noticeable at the higher ends of the continuum. We would expect increased cognitive flexibility and wide mental associations as well at those higher ends.

Although art is a symbolic system of communication unique to humans and ubiquitously present everywhere human groups live, artistic talent does not follow a normal curve within the population. Its rarity makes talent a highly prized commodity. Indeed, as has been suggested, identifying and appreciating artistically talented individuals could have been a critical pivotal feature in the formation and advancement of *Homo sapiens* society (Dissanayake, 1995; Lewis-Williams, 2002). Moreover, inheritance of talent is nonlinear: Highly talented artists, for example, do not necessarily transmit their "talent genes" to their children. Bringing to mind all the talented artists and scientists from the last 200 years alone very rarely brings to mind their progeny, and in such rare cases only children, not grandchildren. The same principle of lack of seemingly direct inheritance applies to creativity.

The special case of artistic autistic savants is relevant here. Talent for drawing and composing realistic spatial depictions is preserved in a tiny fraction of individuals suffering from autism, a severe social communication condition sometimes accompanied by mental retardation. These individuals are known as artistic autistic savants. The remarkable aspect of the condition is that despite the extensive damage in the brain, the nature of which is not completely understood (Minshew & Keller, 2010), islands of drawing talent are preserved. Yet, very little creativity is exhibited in such cases. By and large, instruction and teaching do not help the individual go beyond his or her talent, unlike normal talented artists who improve, change, and benefit from instruction (Sacks, 1995, 2004).

Lessons from Brain Damage: Creativity in Professional Artists with Brain Damage

The brain underpinning of our psychological capacities is traditionally inferred from consequences of damage to the brain. The fields of neuropsychology and neurology are dedicated to uncovering the behavioral consequences of brain damage. For example, damage to Broca's area, which lies in the inferior portion of the left frontal lobe, toward the lobe's posterior, leads to severe speaking difficulties and some comprehension problems of sentences laden with specific grammatical constructions (specifically, conjunctions). In contrast, damage to Wernicke's area, which lies in the left temporal lobe in the posterior region of the auditory region, leads to severe deficits in comprehension of spoken and written language. Damage to the right inferior region spanning the temporal and occipital lobes can lead to profound difficulties in recognizing previously familiar faces. These are all examples of what is known as functional localization in the brain. With regard to creativity and its brain localization, we do not yet have enough knowledge to pinpoint it.

However, visual and musical artists who have practiced their craft for many years prior to the neurological event that caused the brain damage can help reveal and point the way to the neuroanatomical underpinnings of creativity (Viskontas & Miller, this vol.; Zaidel, 2009, 2010). The damage disrupts normal neuronal connectivity, and the effects of the disruption can be measured in behavioral performance by the patient. By studying their artistic output following the event, we can hone in on the relevant brain regions and pathways (Bogousslavsky & Boller, 2005; Rose, 2004; Zaidel, 2005). With localized brain damage due to stroke, tumor, or surgical tissue excision, the behavior exhibited by the patient is attributed to neuronal disruption of the localized area. By contrast, when the damage is diffuse throughout the brain, as is the case in various dementing conditions such as Alzheimer's disease (AD) or frontotemporal dementia (and other such diseases), large areas of the brain are affected. With the latter kind of case, it is difficult to attribute the consequences of the damage to a specific region and its specialized neuronal attributes.

Compared to the incidence of brain damage in the rest of the population, the number of documented such cases of artists is very small. Critically, examination of their postdamage output has revealed that on the whole they (1) continue to produce art, sometimes prolifically, and (2) their creativity does not diminish (Zaidel, 2005, 2009, 2010). Importantly, artists with AD produce art well into the disease, even as the condition worsens, with no visible reduction in their creativity (e.g., Fornazzari, 2005). Interestingly, a study of aesthetic preference in AD patients found no statistically significant difference in aesthetic preferences between the patients and normal control subjects (Halpern, Ly, Elkin-Frankston, & O'Connor, 2008). However, they do cease to produce art when severe motoric deficits develop and profoundly restrict their hand movements.

Moreover, patients with frontotemporal dementia do not become more creative following the onset of the disease (Rankin et al., 2007). A recent study of seventeen patients (nonartists) suffering from a frontal variant of frontotemporal dementia revealed that they have poor and diminished creativity (de Souza et al., 2010). When artistic talent is observed for the first time with such patients (e.g., Mell, Howard, & Miller, 2003; Miller, Boone, Cummings, Read, & Mishkin, 2000), creativity is not the source of the artistic production; rather, the removal of inhibition due to damage to the frontal lobes is the cause. That is, the productions in these patients are not necessarily creative but rather can be interpreted to be displays of previously repressed, latent artistic talent. As with all lessons from brain damage, what we learn from such cases is that in the normal intact brain, the frontal lobes play a role in creativity.

The importance of studying visual artists with brain damage is that light is shed on the relationship between art and brain through examination of their postdamage works compared to their predamage output. The key questions concern any alterations in creativity or loss of talent or skill. Approximately 45 to 55 cases with unilateral damage or with diffuse damage have been described thus far in the neurological literature, and a review of the majority of these cases indicates that on the whole artists go on producing art despite the damage's laterality or localization (Zaidel, 2005). So, together, these results suggest that artistic creativity, or talent and skill, as are ideas, concepts, and symbolic cognition, are generally diffusely represented in the brain; no single "center," region, or pathway controls art-related creativity, cognition, and production.

Furthermore, no specific technique or style alterations are associated with localization of the damage, or its etiology. Artists adhering to the abstract art genre (style) adhere to it following brain damage, and the same is true of the realistic style pre- and postdamage. This implies that the neurological foundations of genre, too, are diffusely represented, and through redundancy of functional representation they survive regional damage. Some brain-damaged artists develop techniques to compensate for loss of perceptual and cognitive specialization. However, these techniques are subtle and too complex to group into coherent categories.

Thus far, any seemingly art-related alterations following brain damage in established professional artists can be explained by general defects in perceptual and cognitive processing that are observed in nonartists suffering from similar brain damage. One example is loss of accurate depictions of three-dimensional space (3D) with right parietal lobe damage (De Renzi, 1982). Hemi-neglect or hemi-inattention of the left half of space is another example. This condition typically occurs following right parietal lobe damage, and its manifestation in visual artists is lack of completion of the left half of the canvas. In the majority of cases, however, neglect symptoms are short lived, lasting approximately six weeks or so. The presence of the neglect syndrome has been attributed to imbalance created by the damage between intact and diseased tissue (Zaidel, 2005), as well as to an abnormal control of the healthy tissue in the left hemisphere over the right half of space (i.e., the space that is not ignored) (Kinsbourne, 1977). In sum, deficits in perceptual deficits can be found in both artists and nonartists and when observed in the latter do not inform us of art-related neural substrates. Thus, the cognitive functions specialized in both cerebral hemispheres should be regarded as being involved in the whole artistic process.

What can we glean from neurological cases of individuals who have *not* practiced art prior to suffering brain damage, and commence to practice art after the damage? To begin with, producing art per se is not necessarily creative. The production is a reflection of many things, including personal wishes and thoughts, talent, skills, and intelligence. Turning to art following the damage can be explained in terms of alternatives to lost communicative cognitive functions so that the art becomes a substitute for the loss of previously used communication modes (e.g., speaking, writing). With art, particularly drawing and painting, a new method of communication is established. Published illustrations of such productions do not necessarily bespeak of creativity (Pollak, Mulvenna, & Lythgoe, 2007). Moreover, judging from the visual details and quantity of works, the current thinking is that art production by these neurological cases has a strong obsessive-compulsive feature (Chatterjee, 2006; Finkelstein, Vardi, & Hod, 1991; Lythgoe, Polak, Kalmus, de Haan, & Khean, 2005). This, in turn, implies an interaction between the neuronal underpinning of the obsessive-compulsive disorder and some types of art expression.

It should be emphasized that acquired damage to the right or left hemisphere does not lead to disappearance, abolishment, or elimination of artistic creativity. The damage does not prevent continuation of art production regardless of its etiology, laterality, or extent (Lakke, 1995). This suggests a wide and diffuse representation of the function of art creativity in the brain, as well as talent and skill. Several spared neuronal circuitries seem to go on functioning despite the presence of damage. The clue to brain and creativity, then, might lie in the actions of neurotransmitters. The following section addresses this issue.

Neurotransmitters and Artists with Parkinson's Disease

Artists suffering from Parkinson's disease (PD) can help shed further light on the issues under discussion here. The disease is characterized by tremors and severe depletion of dopamine, and as the disease progresses, motor coordination becomes severely compromised. Treatment consists in medically increasing levels of the neurotransmitter dopamine. Lakke (1999) reported that despite suffering from PD, the artists in his studies continued to produce visual art, and this despite having a tremor in their dominant hand. Some adjusted their art by utilizing the advantages conferred by the tremor. Later, reports of PD patients, of both artists and nonartists, linked dopamine agonist medication to artistic output and to a strong obsessive-compulsive component in the art production. Schrag and Trimble (2001) describe the case of a talented PD patient who began to write high-quality poems within the first month after initiation of dopamine agonist medication (lisuride as well as levodopa), although he had not written poetry previously. He wrote ten poems in the first year. His productivity went uninterrupted and eventually he won an important poetry prize. It should be mentioned that his grandfather on his mother's side was an accomplished poet. Schrag and Trimble suggest that the poetry writing could be due to the loss of inhibition (because of frontal lobe damage), which could have facilitated the literary productivity, as well as the stimulation induced by the neurotransmitters dopamine and serotonin. The implication is that an alteration in neurotransmitter balance in the brain together with specific neuroanatomical brain alterations can contribute to enhanced artistic creativity. Obviously, artistic talent has to be in place (in the brain) to begin with, or else no amount of disinhibition, frontal lobe damage, or neurotransmitter imbalance would help artistically.

Walker and colleagues (Walker, Warwick, & Cercy, 2006) reported on the enhancement of productivity in a case of a talented visual artist who, following initiation of dopamine medication, increased his drawing and sketching activity. Similarly, Kulisevsky and colleagues (2009) presented an amateur PD artist in whom enhanced dopamine medication resulted in increased painting activity plus a change in personal artistic technique. Schwingenschuh and colleagues (2010) discussed four successful artistic PD patients (a playwright, a fiction writer, and two professional painters) who, after receiving dopamine agonist medication, engaged in compulsive artistic productive output. All of these published reports are of cases of individuals who prior to their disease onset were talented, practicing artists. The disease condition did not obliterate their talent or creativity. The specific medication of dopamine replacement contributed to enhanced productivity of a compulsive nature (Chatterjee, Hamilton, & Amorapanth, 2006). The interplay of the frontotemporal lobes and dopamine has been emphasized by Flaherty's (2005) study, although, as mentioned above, frontotemporal dementia patients do not become more creative following the onset of the disease (Rankin et al., 2007; de Souza et al., 2010).

By inference to the normal brain, the foregoing suggests that the dopaminergic system and frontal lobe regions are indeed involved in positive creativity. Recently, de Manzano and associates (2010) suggested that the D2 receptor in the dopaminergic signaling system, particularly in the thalamus, plays an important role in creativity of healthy, psychiatrically free individuals. The frontal lobes play a major role in planning ahead, working memory, and cognitive flexibility, and the thalamus is an important relay station in the brain; these are features of the mind that contribute to rich forms of creativity.

The threshold for the effects of dopamine in the creative process is unknown at the present time. The density of D2 receptors could explain individual variability in creativity, for example. Dopamine is a neurotransmitter that is involved in widely varied human behavior, including sensations of pleasure, impulse control problems, drug addiction, and gambling (Flaherty, 2005). Whether or not dopamine in conjunction with intact functioning of several brain regions contributes to remarkable creativity needs to be answered by future research.

Disinhibition, Neuronal Circuitry, and Neuronal Connectivity

As noted in the introductory paragraph to this chapter, creativity consists of transcending the given, accepted, and common knowledge. We have known for a long time that when there is damage to the frontal lobes, patients exhibit behavioral disinhibition: they use curse words, engage in inappropriate behavior, wear unkempt clothing when out in public, and engage in reckless activities (Fuster, 2001). Does this mean that patients with frontal lobe damage are creative? Not necessarily. Indeed, according to a recent paper, they become anything but creative (de Souza et al., 2010). The frontal lobes have rich connections to the rest of the brain, including regions that are critical for memory, concept formation, and problem solving. We can, however, obtain insights into the brain's underpinnings of creativity from research on decision making particularly as related to overcoming inhibition. Aron and associates (2007) have implicated the prefrontal cortex and the basal-ganglia network in overcoming inhibitory neural circuitries, ones that impose inhibition on impulsive behavior. In decision-making research, what is of interest is what happens in the brain when people overcome the status quo knowledge. Creativity, after all, is the process of introducing something new, something over and above the given and the known. Fleming and associates (2010) asked subjects to detect visual stimuli in a paradigm that varied the difficulty level while they were being scanned with fMRI. They found increased activity in the subthalamic nucleus when the status quo was overcome with a difficult decision. Furthermore, their data analysis confirmed a neuronal circuitry involving the prefrontal cortex and basal ganglia in status quo rejection. Together, everything else being equal, this pathway may be involved in transcending the given, established status quo of knowledge. By inference, then, the pathway would be involved in the creative process.

Creativity in Artists Is Prescient

Unbound by rigid rules such as those imposed in scientific investigations, successful artists are free to let their minds soar, and this, in combination with their talent and intelligence, enables some of them to experiment and produce highly original works (Miller, 2005). They are unbound by the cognitive associations required by highly detailed scientific knowledge. Consequently, artists have greater freedom in expressing and exploring the limits of creativity. Ultimately, art in all its formats—literature, poetry, music, painting, film, sculpture, dance, theater, or photography—reflects the mind of the artist, whether he or she has a normal brain or is an autistic savant, a person with frontotemporal dementia, a patient who had unilateral stroke, or an exceptional artist such as Monet, Picasso, or Modigliani. The artist's studio, whatever and wherever it is, is where the workings of the creative mind are continuously tested. Indeed, visual artists have often inspired scientists to view their research projects with a fresh and creative perspective (Miller, 2000).

All of this suggests that artistic creativity permits infinite combinatorial possibilities. In a speculative vein, it may be that the imperfect, unbalanced display of creativity is precisely what creativity is: a process that computes deviations and incongruities in the normal pattern of neuronal activity. The underlying neural circuitry, electrical and chemical, appears to be nonlinear. And, importantly, as mentioned earlier, creativity is not unique to art; we see it in science, technology, business, politics, and all around us.

Conclusion

Creativity and innovation are not unique to humans. Such behaviors in animals (birds, monkeys, apes) suggest that nonhuman brain mechanisms are fine-tuned to deal with experience in innovative ways, that the roots of creativity originated in the human biological ancestry. With humans, these ways are expressed in creativity in the arts, science, business, technology, and daily life.

The fact that established, professional artists suffering from damage to different cortical regions nevertheless continue to produce art with retained creativity implies the absence of a single neuronal circuitry for creativity.

The inseparable interaction of artistic skill and talent from the factor of creativity adds to the complexity of the issue. We have also seen that neurochemical factors play a role in human creativity, as is intelligence and its genetic inheritance. Dopamine has recently been implicated with creativity, and this is an important neuroscientific lead in the quest. So far, psychometric tests have been used in this context of the dopamine; predefined laboratory tests, however, do not necessarily characterize spontaneous creativity such as "eureka" moments or the type of creativity displayed by established artists and scientists. A complex interplay of neural factors contributes to creativity, and its components are yet to be deciphered. It would be neat and convenient if we could pinpoint the process that gives rise to creativity, but the brain—as many other biological, chemical, and physical systems—does not follow regular, orderly rules. In sum, it may be that the imperfect, unbalanced display of human creativity is precisely what creativity is: a particular yet irregular neuronal process reflecting deviations in the steady pattern of neuronal activity.

References

Aron, A. R., Durston, S., Eagle, D. M., Logan, G. D., Stinear, C. M., & Stuphorn, V. (2007). Converging evidence for a fronto-basal-ganglia network for inhibitory control of action and cognition. *Journal of Neuroscience, 27*, 11860–11864.

Benson-Amram, S., & Holekamp, K. E. (2012). Innovative problem solving by wild spotted hyenas. *Proceedings of the Royal Society of London, Series B: Biological Sciences, 279*, 4087–4095.

Bogousslavsky, J., & Boller, F. (Eds.). (2005). *Neurological disorders in famous artists: Frontiers in neurological neuroscience*. Basel: Karger.

Bonner, J. T. (1980). *The evolution of culture in animals*. Princeton: Princeton University Press.

Byrne, R. W. (2003). Novelty in deception. In K. N. Laland & S. M. Reader (Eds.), *Animal innovation* (pp. 237–259). Oxford: Oxford University Press.

Byrne, R. W., & Bates, L. A. (2010). Primate social cognition: Uniquely primate, uniquely social, or just unique? *Neuron, 65*, 815–830.

Byrne, R. W., & Whiten, A. (1990). Tactical deception in primates: The 1990 database. *Primate Report, 27*, 1–101.

Chatterjee, A. (2006). The neuropsychology of visual art: Conferring capacity. *International Review of Neurobiology, 74*, 39–49.

Chatterjee, A., Hamilton, R. H., & Amorapanth, P. X. (2006). Art produced by a patient with Parkinson's disease. *Behavioural Neurology, 17,* 105–108.

de Manzano, Ö., Cervenka, S., Karabanov, A., Farde, L., & Ullén, F. (2010). Thinking outside a less intact box: Thalamic dopamine D2 receptor densities are negatively related to psychometric creativity in healthy individuals. *PLoS ONE, 5,* e10670.

De Renzi, E. (1982). *Disorders of space exploration and cognition.* New York: Wiley.

de Souza, L. C., Volle, E., Bertoux, M., Czernecki, V., Funkiewiez, A., Allali, G., et al. (2010). Poor creativity in frontotemporal dementia: A window into the neural bases of the creative mind. *Neuropsychologia, 48,* 3733–3742.

Deary, I. J., Penke, L., & Johnson, W. (2010). The neuroscience of human intelligence differences. *Nature Reviews: Neuroscience, 11,* 201–211.

Dissanayake, E. (1995). *Homo aestheticus: Where art comes from and why.* Seattle: Washington University Press.

Fink, A., Grabner, R. H., Benedek, M., Reishofer, G., Hauswirth, V., Fally, M., et al. (2009). The creative brain: Investigation of brain activity during creative problem solving by means of EEG and fMRI. *Human Brain Mapping, 30,* 734–748.

Finkelstein, Y., Vardi, J., & Hod, I. (1991). Impulsive artistic creativity as a presentation of transient cognitive alterations. *Behavioral Medicine, 17,* 91–94.

Flaherty, A. W. (2005). Frontotemporal and dopaminergic control of idea generation and creative drive. *Journal of Comparative Neurology, 493,* 147–153.

Fleming, S. M., Thomas, C. L., & Dolan, R. J. (2010). Overcoming status quo bias in the human brain. *Proceedings of the National Academy of Sciences of the United States of America, 395,* 1120–1125.

Fornazzari, L. R. (2005). Preserved painting creativity in an artist with Alzheimer's disease. *European Journal of Neurology, 12,* 419–424.

Fuster, J. M. (2001). The prefrontal cortex—an update: Time is of the essence. *Neuron, 30,* 319–333.

Gardner, H. E. (1994). *Creating minds: An anatomy of creativity as seen through the lives of Freud, Einstein, Picasso, Stravinsky, Eliot, Graham, and Gandhi.* New York: Basic Books.

Goodall, J. (1986). *The chimpanzees of Gombe: Patterns of behavior.* Cambridge, MA: Harvard University Press.

Halpern, A. R., Ly, J., Elkin-Frankston, S., & O'Connor, M. G. (2008). "I know what I like": Stability of aesthetic preference in Alzheimer's patients. *Brain and Cognition, 66,* 65–72.

Hinde, R. A., & Fisher, J. (1951). Further observations on the opening of milk bottles by birds. *British Birds, 44,* 393–396.

Hulshoff Pol, H. E., Schnack, H. G., Posthuma, D., Mandl, R. C. W., Baare, W. F., van Oel, C., et al. (2006). Genetic contributions to human brain morphology and intelligence. *Journal of Neuroscience, 26,* 10235–10242.

Jung, R. E., Grazioplene, R., Caprihan, A., Chavez, R. S., & Haier, R. J. (2010). White matter integrity, creativity, and psychopathology: Disentangling constructs with diffusion tensor imaging. *PLoS ONE, 5,* e9818.

Jung, R. E., Segall, J. M., Jeremy Bockholt, H., Flores, R. A., Smith, S. M., Chavez, R. S., et al. (2009). Neuroanatomy of creativity. *Human Brain Mapping, 31,* 398–409.

Kaufman, A. B., Butt, A. E., Kaufman, J. C., Colbert-White, E. N. (2011). Towards a neurobiology of creativity in nonhuman animals. *Journal of Comparative Psychology, 125,* 255–272.

Kawai, M. (1965). Newly acquired pre-culture behavior of a natural troop of Japanese monkeys on Koshima island. *Primates, 6,* 1–30.

Kawamura, S. (1959). The process of sub-culture propagation among Japanese macaques. *Primates, 2,* 43–60.

Kinsbourne, M. (1977). Hemi-neglect and hemisphere rivalry. *Advances in Neurology, 18,* 41–49.

Kulisevsky, J., Pagonabarraga, J., & Martinez-Corral, M. (2009). Changes in artistic style and behaviour in Parkinson's disease: Dopamine and creativity. *Journal of Neurology, 256,* 816–819.

Lakke, J. P. W. F. (1995). The black hole in Charley Toorop's "The Three Generations": A neuro-iconographic analysis and reconstruction. *Clinical Neurology and Neurosurgery, 97,* 269–276.

Lakke, J. P. W. F. (1999). Art and Parkinson's disease. *Advances in Neurology, 80,* 471–479.

Laland, K. N. (2011). Cause and context in the biological sciences. *Behavioral Ecology, 22,* 223–234.

Laland, K. N., Odling-Smee, J., & Myles, S. (2010). How culture shaped the human genome: Bringing genetics and the human sciences together. *Nature Reviews: Genetics, 11,* 137–148.

Laland, K. N., & Reader, S. M. (2010). Comparative perspectives on human innovation. In M. J. O'Brien & S. J. Shennan (Eds.), *Innovation in cultural systems: Contributions from evolutionary anthropology*. Cambridge, MA: MIT Press.

Lefebvre, L., Reader, S. M., & Sol, D. (2004). Brains, innovations, and evolution in birds and primates. *Brain, Behavior, and Evolution, 63*, 233–246.

Lewis-Williams, D. (2002). *The mind in the cave: Consciousness and the origins of art.* London: Thames & Hudson.

Li, J., Yu, C., Li, Y., Liu, B., Liu, Y., Shu, N., et al. (2009). COMT val158met modulates association between brain white matter architecture and IQ. *American Journal of Medical Genetics, Part B: Neuropsychiatric Genetics, 150*, 375–380.

Lythgoe, M. F., Polak, T., Kalmus, M., de Haan, M., & Khean, C. W. (2005). Obsessive, prolific artistic output following subarachnoid hemorrhage. *Neurology, 64*, 397–398.

Mell, C. J., Howard, S. M., & Miller, B. L. (2003). Art and the brain: The influence of frontotemporal dementia on an accomplished artist. *Neurology, 60*, 1707–1710.

Miller, A. I. (2000). *Insights of genius: Imagery and creativity in science and art.* Cambridge, MA: MIT Press.

Miller, A. I. (2005). Creativity special: One culture. *New Scientist, 2523*, 44–45.

Miller, B. L., Boone, K., Cummings, J. L., Read, S. L., & Mishkin, F. (2000). Functional correlates of musical and visual ability in frontotemporal dementia. *British Journal of Psychiatry, 176*, 458–463.

Minshew, N. J., & Keller, T. A. (2010). The nature of brain dysfunction in autism: Functional brain imaging studies. *Current Opinion in Neurology, 23*, 124–130.

Neubauer, A. C., & Fink, A. (2009). Intelligence and neural efficiency: Measures of brain activation versus measures of functional connectivity in the brain. *Intelligence, 37*, 223–229.

Neubauer, A. C., Grabner, R. H., Fink, A., & Neuper, C. (2005). Intelligence and neural efficiency: Further evidence of the influence of task content and sex on the brain-IQ relationship. *Cognitive Brain Research, 25*, 217–225.

Pollak, T. A., Mulvenna, C. M., & Lythgoe, M. F. (2007). De novo artistic behaviour following brain injury. *Frontiers of Neurology and Neuroscience, 22*, 75–88.

Rankin, K. P., Liu, A. A., Howard, S. M., Slama, H., Hou, C. E., Shuster, K., et al. (2007). A case-controlled study of altered visual art production in Alzheimer's and FTLD. *Cognitive and Behavioral Neurology, 20*, 48–61.

Reader, S. M., & Laland, K. N. (2002). Social intelligence, innovation, and enhanced brain size in primates. *Proceedings of the National Academy of Sciences of the United States of America, 99*, 4436–4441.

Rose, F. C. (Ed.). (2004). *Neurology of the arts: Painting, music, literature.* London: Imperial College Press.

Sacks, O. (1995). *An anthropologist on Mars*. New York: Alfred A. Knopf.

Sacks, O. (2004). Autistic geniuses? We're too ready to pathologize. *Nature, 429*, 241.

Schrag, A., & Trimble, M. (2001). Poetic talent unmasked by treatment of Parkinson's disease. *Movement Disorders, 16*, 1175–1176.

Schwingenschuh, P., Katschnig, P., Saurugg, R., Ott, E., & Bhatia, K. P. (2010). Artistic profession: A potential risk factor for dopamine dysregulation syndrome in Parkinson's disease? *Movement Disorders, 10*, 120–124.

Sternberg, R. J. (1997). *Successful intelligence*. New York: Plume.

Sternberg, R. J., & O'Hara, L. A. (1999). Intelligence and creativity. In R. J. Sternberg (Ed.), *Handbook of intelligence* (pp. 252–272). New York: Cambridge University Press.

Taylor, A. H., Miller, R., & Gray, R. D. (2012). New Caledonia crows reason about hidden causal agents. *Proceedings of the National Academy of Sciences of the United States of America, 109*, 16389–16391.

Walker, R. H., Warwick, R., & Cercy, S. P. (2006). Augmentation of artistic productivity in Parkinson's disease. *Movement Disorders, 21*, 285–286.

Wyles, J. S., Kunkel, J. G., & Wilson, A. C. (1983). Birds, behavior, and anatomical evolution. *Proceedings of the National Academy of Sciences of the United States of America, 80*, 4394–4397.

Zaidel, D. W. (2005). *Neuropsychology of art: Neurological, cognitive, and evolutionary perspectives*. Hove: Psychology Press.

Zaidel, D. W. (2009). Brain and art: Neuro-clues from intersection of disciplines. In M. Skov & O. Vartanian (Eds.), *Neuroaesthetics* (pp. 153–170). Amityville, NY: Baywood.

Zaidel, D. W. (2010). Art and brain: Insights from neuropsychology, biology, and evolution. *Journal of Anatomy, 216*, 177–183.

IV Pharmacology and Psychopathology

8 Pharmacological Effects on Creativity

David Q. Beversdorf

In the pursuit of understanding of how creative processes are carried out in the brain, it is particularly important to understand how these processes are affected by the major neurotransmitter systems. Not only does this allow a greater understanding of the mechanism, but as will be discussed below, it allows a greater opportunity for clinical intervention than more anatomical approaches. Most research has focused on the catecholaminergic systems—the dopaminergic system and the noradrenergic system—but evidence is beginning to be explored for other systems as well. A greater volume of literature exists for the pharmacological effects on other executive functions highly interrelated with creativity, such as set-shifting and working memory. These will be discussed as relevant for creativity, but the distinctions between creativity and these other executive functions may also be quite critical, as some evidence already suggests. Studies on the effects of pharmacology on creativity have generally focused on creativity in problem-solving (convergent) tasks and creativity in divergent tasks.

The Dopaminergic System

Early suggestions as to the potential role of the dopaminergic system in creativity came from the effects on the semantic network as observed in priming studies, since the ability to search within the semantic network is a critical component of both semantic priming and verbal creativity tasks. In 1996, Kischka et al. demonstrated in a priming experiment in healthy individuals that word recognition occurs more rapidly when presented 700 milliseconds after exposure to another directly related or indirectly related word. However, after administration of L-dopa, the precursor for dopamine, only words presented after directly related words are recognized quickly. Kischka et al. proposed a role of the dopaminergic system in restriction of the semantic network in priming. Spreading activation of either a directly or indirectly related word facilitated word recognition without L-dopa, but only the directly related word facilitated word recognition with L-dopa. This effect appears to be sensitive to the time between the initial and target stimuli, likely a reflection of the effects of the timing of spreading activation. Subsequent research by Angwin et al. (2004) demonstrated that L-dopa affected both direct and indirect priming with an interstimulus interval of 500 milliseconds, but had no affect at 250 milliseconds. This would seem consistent with what might be expected with an effect on a widely distributed network. However, since L-dopa is a dopamine precursor, it remained unclear as to which specific dopamine receptors might be responsible for the priming effect. Studies in healthy volunteers (Roesch-Ely et al., 2006), in addition to studies in patients with Parkinson's disease (Pederzolli et al., 2008), suggest that the priming effect is mediated by action on the D1 receptor.

In order to begin to examine how dopaminergic agents might affect semantic networks, researchers looked at the effect of L-dopa on functional connectivity during functional magnetic resonance imaging (fcMRI) using a nonpriming language task: a word categorization task. An isolated increase in connectivity was observed with L-dopa between the left fusiform and the receptive language areas, with no other region pairs affected (Tivarus et al., 2008). Since the left fusiform is considered the visual word form receptive area (Beversdorf et al., 1997), this would appear to fit with the effects on priming, as the interaction between this fusiform area (critical for visual word form recognition) and Wernicke's area (critical for processing word meaning) would be essential for priming effects. However, since the predominant target among cortical areas for dopaminergic projecting fibers is the frontal lobe (Hall et al., 1994; Lidow et al., 1991) (figure 8.1), such an effect of L-dopa on these posterior regions seems unexpected. Subsequent evidence using independent component analysis of functional magnetic resonance imaging (fMRI) data during language tasks suggests that the posterior effects of L-dopa may be mediated indirectly by the frontothalamic connections from the areas containing the frontal projections of the dopaminergic fibers (Kim et al., 2010). A recent fMRI study examining the effect of L-dopa during priming revealed changes in region-of-interest (ROI) activation with drug in the dorsal prefrontal cortex, anterior cingulate, left rolandic operculum, and left middle temporal gyrus (Copland et al., 2009), which also may suggest an indirect frontal-posterior interaction.

This literature surrounding the relationship between the catecholaminergic systems and semantic priming has led to a proposal of its role in

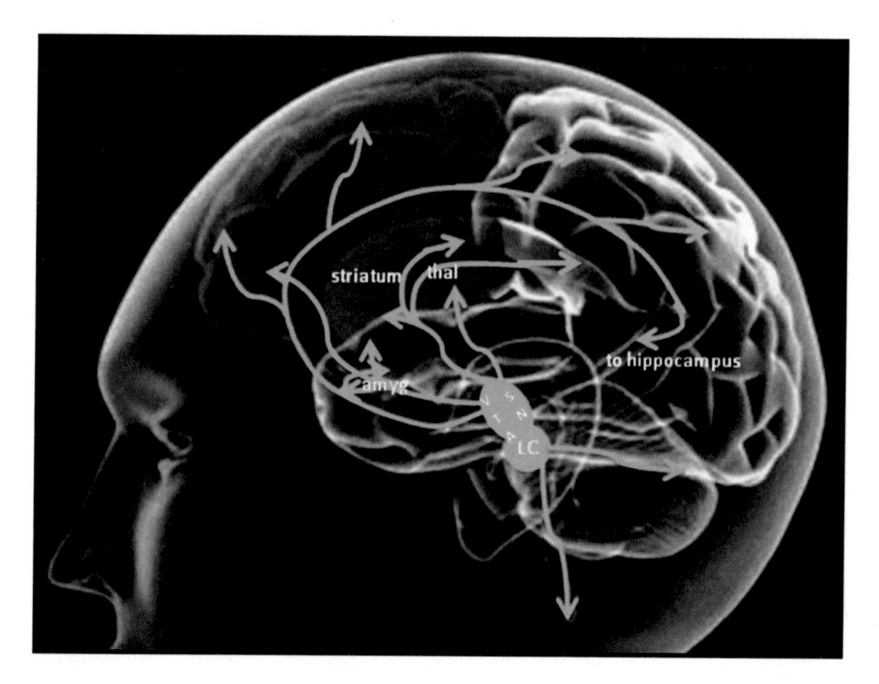

Figure 8.1
Noradrenergic (blue) and dopaminergic (green) pathways. In blue: The locus coeruleus (LC) projects posteriorly to the cerebellum and up to the thalamus (thal) and amygdala (amyg), as well as throughout the neocortex along a pericingular tract, also terminating posteriorly at the hippocampus (Heimer, 1995). The descending fibers to the spinal cord are also shown. Not shown is the lateral tegmental noradrenergic system which also projects to the amygdala and down to the spinal cord. In green: Projections from the substantia nigra (SN) to the striatum are demonstrated, as are projections from the ventral tegmental area (VTA) to the amygdala (amyg), ventral striatum, and frontal cortex (Heimer, 1995). Not shown are the tuberoinfundibular and posterior hypothalamic dopaminergic systems.

verbal creativity as well as potentially in nonverbal forms of creativity (Heilman et al., 2003). However, in consideration of the research based on administration of L-dopa, it must be noted that L-dopa is also a precursor to norepinephrine. Further study has been initiated in hopes of disentangling the potential effects of the dopaminergic and noradrenergic systems on priming and creativity in problem solving, both of which are known to be sensitive to the action of catecholaminergic agents (dopamine and norepinephrine) on semantic networks (Campbell et al., 2008; Kischka et al., 1996). Dopaminergic agonists were found to have no effect on creativity in verbal problem solving (Smyth & Beversdorf, 2007), and noradrenergic agents did not appear to affect priming in the manner observed with dopaminergic agents (Cios et al., 2009). This appears to suggest a role for the dopaminergic system (but not the noradrenergic system) in effects on automatic searches of the semantic network as with word recognition (Cios et al., 2009; Kischka et al., 1996), and a role for the noradrenergic system (but not the dopaminergic system) in effects on controlled searches of the semantic network as with verbal problem solving (Campbell et al., 2008; Smyth & Beversdorf, 2007). The relationship between the noradrenergic system and creativity will be discussed subsequently. However, despite these findings, other recent research has suggested a more direct relationship between the dopaminergic system and creativity. Studies examining rate of eyeblink, proposed as a marker of dopaminergic activity, demonstrated an inverted U-shaped relationship between eyeblink rate and creativity as assessed by an alternate uses task (AUT) and the remote associates task (RAT) (Chermahini & Hommel, 2010). Genetic studies demonstrate a relationship between D2 receptor polymorphisms and a composite creativity score as well as performance on verbal creativity, as assessed by object use fluency and sentence fluency from three words (Reuter et al., 2006). A positive association has been found in the relationship between gray matter volume in dopaminergic subcortical regions as well as the right dorsolateral prefrontal cortex and divergent thinking (fluency for unusual uses and unimaginable things) with voxel-based morphometry on MRI (Takeuchi et al., 2010), and a negative association in the relationship between thalamic D2 receptor densities and performance on verbal, figural, and numerical fluency tasks with receptor binding studies using positron emission tomography (de Manzano et al., 2010). A case study describing changes in artistic behavior with dopaminergic agonists in Parkinson's disease has also been proposed as evidence for a relationship between the dopaminergic system and creativity (Kulisevsky et al., 2009), but the effects on interest in (as well as obsession with) artistic output and effects on style in such cases are hard to disentangle from other aspects of creativity (Chatterjee et al., 2006). While this array of indirect supportive data for a role for the dopaminergic system in creativity is of interest, the distinction between the roles of the noradrenergic and dopaminergic systems in creativity is in need of further study.

The dopaminergic system has a range of other cognitive effects, including other aspects of executive function closely related to creativity, in addition to the well-known effects on the motor system. Research in animal models has demonstrated varying effects of dopaminergic agents on set-shifting tasks, differing according to which receptor subtype each agent impacts (Floresco et al., 2005). Among set-shifting tasks, this effect appears to be specific to intradimensional set shifting (Robbins, 2007). Whereas agonists for both D1 and D2 receptors did not affect set shifting, D2 antagonists impaired set shifting in rodents (Floresco et al., 2006; Stefani & Moghaddam, 2005), an effect also observed in humans (Mehta et al., 2004). In further support of a role of the dopaminergic system in executive function, ability to maintain and flexibly alter cognitive representations in response to environmental demands is known to be impaired in Parkinson's disease (Cools, 2006). Computational models propose that phasic stimulation of D2 receptors in the striatum drives flexible adaptation of cognitive representations that are maintained by the prefrontal cortex (Cohen et al., 2002), which contrasts with the effect on priming, which appears to be mediated by D1 receptors (Roesch-Ely et al., 2006; Pederzolli et al., 2008). Receptor specificity of effects on creativity is not known. It should be noted, regarding the effect of dopamine on executive function in patient populations, that the interaction between dopaminergic agonists and Parkinson's disease and their effect on cognition is complex for set shifting as well as working memory. Early in Parkinson's disease, greater dopaminergic depletion in the dorsal striatum leads to impaired adaptation in responses and updating in working memory, which is improved by L-dopa, while maintenance of working memory is less affected. However, L-dopa also can excessively enhance reward biases due to effects on the relatively intact ventral striatum (Cools, 2006). These other cognitive effects of the dopaminergic system are discussed further below.

Regarding these dopaminergic effects on other closely related executive functions, in healthy subjects, individuals with lower working memory capacity tend to be the ones who benefit from increased prefrontal function with dopaminergic stimulation (Gibbs & D'Esposito, 2005; Kimberg et al., 1997). This is likely related to the fact that dopamine synthesis capacity in the striatum is related to working memory capacity, such that those with the least working memory capacity also have less dopamine, and therefore benefit from dopaminegic stimulation (Cools, Gibbs et al., 2008), again suggesting an inverted U-shaped relationship between performance and dopaminergic function. In animal models, this effect on working memory appears to be mediated by action at the D1 receptor (Arnsten et al., 1994; Sawaguchi & Goldman-Rakic, 1991; Williams and Goldman-Rakic, 1995). Dopamine also appears to be critical for a range of other aspects of cognition involving frontal-subcortical circuits, including the temporal coupling of deliberation and execution during decision making, as dopamine replacement reverses the delay specific to decision-related hesitations, independent of motor slowing, in situations requiring decision making in uncertainty in patients with Parkinson's disease (Pessiglione et al., 2005).

Whereas working memory, set shifting ("constrained cognitive flexibility"), and creativity involved in problem solving and divergent thinking tasks (both considered as "unconstrained cognitive flexibility") are highly interrelated, significant further work is necessary to disentangle the roles of the various neurotransmitter systems on each domain. This becomes particularly critical for the effects of the noradrenergic system, as is discussed below.

Another critical role of dopamine has recently become apparent with the development of pathological gambling in the setting of treatment with dopaminergic agonists (Dodd et al., 2005; Gallagher et al., 2007). This has contributed to a greater understanding of the roles of dopamine in decision making, revealing that dopamine neurons encode the difference between expected and received rewards, and interact with other neurotransmitter systems to regulate such decision making (Nakamura et al., 2008). The relationship between this effect of dopamine and creativity is also in need of further exploration.

The Noradrenergic System

The effects of performance anxiety and test anxiety have been associated with activation of the noradrenergic system, leading to the development of treatment limiting the adrenergic activating effects of stress. Propranolol, a centrally acting β-adrenergic antagonist, has long been used for stress-induced impairment in performance on tasks including public speaking in anxiety-prone individuals (Lader, 1988; Laverdue & Boulenger, 1991). This has led to further exploration of the role of the noradrenergic system in the impact of stress on cognitive performance. Furthermore, research involving healthy adolescents with a history of stress-induced cognitive impairment during exams has demonstrated that treatment with the beta-adrenergic antagonist propranolol significantly improved scores on the Scholastic Aptitude Test (SAT) (Faigel, 1991). The effects of stress and the noradrenergic system on cognition are not limited to patients with known stress-induced cognitive impairment. Stress has long been known to impair performance on tasks requiring creativity in healthy individuals (Martindale & Greenough, 1973), and stress is also known to increase activity of the noradrenergic system (Kvetnansky et al., 1998; Ward et al., 1983). In more recent work beginning to examine the effects of stress in individuals without any history of anxiety-related disorders, administration of a well-characterized social stressor characterized by public speaking and mental arithmetic (Kirschbaum et al., 1993) resulted in impaired performance on creative verbal problem solving requiring flexibility of access to lexical, semantic, and associative networks, and the impairment was reversed by propranolol (Alexander et al., 2007). Therefore, the pharmacological and stress effects on cognition in this setting appear to represent a fundamental aspect of the brain-behavior relationship, not requiring the presence of an anxiety-related disorder or a dysregulated noradrenergic system. However, it should be noted that the effect of propranolol in this study does not exclusively implicate the noradrenergic system, since propranolol has also been shown to block the corticosterone-induced impairment of working memory (Roozendaal et al., 2004).

The locus coeruleus contains the majority of neurons in the central nervous system, sending efferents throughout the brain (Barnes & Pompeiano, 1991), without the degree of selectivity of projection to the frontal lobe as is observed with dopamine (figure 8.1). Therefore, an effect of the noradrenergic system on such a distributed function such as creativity might be expected. However, the effects of the noradrenergic system outside the setting of stress are more dependent on the situation. In our previous work, performance on the anagram task was better after administration of the centrally and peripherally acting beta-adrenergic antagonist propranolol than after the noradrenergic agonist ephedrine (Beversdorf et al., 1999; Heilman et al., 2003). Performance on the anagram task was also better after administration of propranolol than after the peripheral-only beta-adrenergic antagonist nadolol (Beversdorf et al., 2002), suggesting that the effect of propranolol on this aspect of cognition is mediated centrally rather than as a result of peripheral feedback. A central mechanism would be predicted by the effect of norepinephrine on the signal-to-noise ratio of neuronal activity within the cerebral cortex (Hasselmo et al., 1997) as well as the correlation between the electronic coupling of noradrenergic neurons in the monkey cortex and proportions of goal-directed versus exploratory behavior (Usher et al., 1999). However, in each of the anagram studies, whereas performance on propranolol was better than on ephedrine or nadolol, it did not significantly differ from placebo (Beversdorf et al., 1999, 2002). In order to better understand the effect of propranolol, subsequent research examined how task difficulty might relate to the drug's effect, since a drug proposed to benefit a broad search of a network due to signal-to-noise effects might be expected to yield a greater beneficial effect when problems are more challenging. Consistent with this, propranolol was found to be beneficial for a range of verbal problem-solving tasks requiring network flexibility when the subject was struggling, and did not confer benefit and in some cases impaired performance when the subject was solving problems with ease (Campbell et al., 2008). The benefit was seen both for the subjects who had the greatest difficulty solving the problems, and for the most difficult problems across all subjects (Campbell et al., 2008). However, propranolol can benefit performance on such language tasks for the easiest problems in situations where there is up-regulated activity of the noradrenergic system due to cocaine withdrawal (Kelley et al., 2005, 2007) and psychosocial stress (Alexander et al., 2007), or where there is anatomic rigidity of the language network (Beversdorf et al., 2007a) due to conditions such as autism (convergent task effects—Beversdorf et al., 2008; divergent task effects—Beversdorf et al., 2011) and Broca's aphasia due to stroke (Beversdorf et al., 2007b).

This variability in effect of noradrenergic drugs between patient groups, as observed in cocaine withdrawal, autism, and aphasia, may also be important for attention deficit disorder. As with the dopaminergic system, early theories proposed that arousal and optimal performance might be related on an inverted U-shaped curve (Yerkes & Dodson, 1908), suggesting such a relationship for the noradrenergic system. Therefore, whereas markedly increased arousal or noradrenergic tone might result in hyper-arousal and inability to perform a task in most individuals, a person with attention deficit disorder might be at baseline at a suboptimal point on the inverted U-shaped curve and require stimulants to perform optimally. Animal data suggest that there is an optimal point of tonic activity of the locus coeruleus that tends to support the emergence of phasic activity, which is associated with focused or selective attention (Aston-Jones et al., 1999; Aston-Jones & Cohen, 2005). Noradrenergic transmission is known to be genetically weaker in some patients with attention deficit disorder (Arnsten, 2007). The effects of drugs on creativity in this population warrant further study. However, preliminary evidence suggests that the effect of stimulants on creativity is limited (Farah et al., 2009).

Variation in the effects of the noradrenergic system is not limited to differences between patient groups. Variation in response can also be observed in unaffected subjects due to exposure to specific environmental conditions. Performance on creative verbal problem-solving tasks is affected by alterations in noradrenergic tone due to changes in posture (Lipnicki & Byrne, 2005), sleep phase (Stickgold et al., 2001), and vagal nerve stimulation (Ghacibeh et al., 2006). These effects appear to be specific to the noradrenergic system and not due to general anxiolytic effects, since such cognitive effects do not appear to occur with non-adrenergic anxiolytics (Silver et al., 2004).

Data are only beginning to emerge regarding how propranolol might affect network performance. Evidence from models derived from data from activity in brain-slice preparations support an effect of norepinephrine on the signal-to-noise ratio of neuronal activity within the cerebral cortex (Hasselmo et al., 1997). Presumably, propranolol increases access to "noise" which in this case would be represented by increased associational input that might be adaptive for solving more difficult problems, where the most immediate response is not optimal (Alexander et al., 2007). In one population characterized by decreased flexibility of network access (Beversdorf et al., 2007a), a potential imaging marker is observed. Decreased functional connectivity on fMRI (fcMRI), or a decrease in the synchrony of activation between activated brain regions, is observed in autism (Just et al., 2004, 2007), believed to be related to the underconnectivity between distant cortical regions in autism (Belmonte et al., 2004). Recent evidence suggests that propranolol increases functional connectivity on fMRI in autism, lending some support to the proposed mechanism of action of propranolol on network access (Narayanan et al., 2010). It is not clear whether the noradrenergic system is dysregulated in autism (Martchek et al., 2006; Minderaa et al., 1994). However, some have proposed that the behavioral effects of fever in autism (Curran et al., 2007) may be related to normalization of a developmentally dysregulated noradrenergic system in autism (Mehler & Purpura, 2009). Regardless of the ambient activity of the noradrenergic system in autism, network rigidity in autism (Beversdorf et al., 2007a) and the suggested effect or propranolol on network access (Campbell et al., 2008) suggest a potential for benefit from noradrenergic agents in autism. Furthermore, case series have suggested a benefit in both social and language domains in autism with beta-adrenergic antagonists (Ratey et al., 1987). The effect of propranolol on task performance has not yet been assessed in imaging studies, as the previous imaging study assessed fcMRI during a task where all subjects perform at ceiling (Narayanan et al., 2010). Decreased functional connectivity has also been observed in patients in acute cocaine withdrawal (Narayanan et al., 2012).

The role of the noradrenergic system in behavior is, of course, not limited to effects on network access and creativity as is detailed above. The noradrenergic system is critical in arousal (Coull et al., 1997, 2004; Smith & Nutt, 1996). Furthermore, the prefrontal cortex, important in a range of types of cognitive flexibility (Duncan et al., 1995; Eslinger & Grattan, 1993; Karnath & Wallesch, 1992; Robbins, 2007; Vilkki, 1992), has afferent projections TO the locus coeruleus in primates (Arnsten & Goldman-Rakic, 1984), containing the majority of noradrenergic neurons in the central nervous system and sending the previously described efferents throughout the brain (Barnes & Pompeiano, 1991). The cognitive flexibility as assessed by verbal problem-solving tasks, such as anagrams and the compound remote associates task (Bowden & Jung-Beeman, 2003), involves a search through a wide network in order to identify a solution ("unconstrained flexibility"), and appears to be modulated by the noradrenergic system as described above, where performance generally improves with decreased noradrenergic activity. Other cognitive flexibility tasks such as the Wisconsin Card Sort Test (Heaton, 1981) involve set shifting between a limited range of options ("constrained flexibility"), which may not be modulated by the noradrenergic system in the same manner, and may even benefit from increased noradrenergic activity (Aston-Jones & Cohen, 2005; Usher et al., 1999). Evidence suggests that decreased noradrenergic activity appears to benefit tasks such as anagrams when subjects are struggling or challenged by stressors (Alexander et al., 2007; Campbell et al., 2008), whereas increased set switching on a two alternative forced choice task is associated with increased noradrenergic tone in primate studies (Aston-Jones & Cohen, 2005; Usher et al., 1999). "Constrained" flexibility can be further subdivided into intradimensional and extradimensional set shifting (Robbins, 2007). The dopaminergic system appears to affect intradimensional set shifting (Robbins, 2007), while the noradrenergic system, specifically by action on the alpha-1 receptor, appears to modulate performance on extradimensional set shifting (Lapiz & Morilak, 2006; Robbins, 2007). The beta-adrenergic receptors in the noradrenergic system, though, appear to modulate the "unconstrained" flexibility (Alexander et al., 2007; Beversdorf et al., 1999, 2002). As discussed with the dopaminergic system, a systematic exploration of the effects of the noradrenergic system on intradimensional and extradimensional set shifting as well as creative problem solving and divergent tasks with "unconstrained cognitive flexibility" is warranted. Noradrenergic agents are also known to have a range of other cognitive effects, including effects on motor learning (Foster et al., 2006), response inhibition (Chamberlain et al., 2006b), working memory, and emotional memory (Chamberlain et al., 2006a).

The role of the noradrenergic system in emotional memory deserves particular comment, due to a potentially important clinical role. Centrally acting beta-adrenergic receptor antagonists are known to reduce the enhancement of memory resulting from emotional arousal (Cahill et al., 1994; van Stegeren et al., 1998). This may contribute to the development of intrusive memories in clinical conditions such as posttraumatic stress disorder (Ehlers et al., 2002, 2004; Smith & Beversdorf, 2008). Evidence is beginning to suggest a secondary preventative role of propranolol in the development of posttraumatic stress disorder, by interfering with reconsolidation (Pitman et al., 2002; Vaiva et al., 2003). Alpha-1 antagonists have similarly revealed benefit in patients with posttraumatic stress disorder (Arnsten, 2007). The relationship between creativity and the emotional effects of the noradrenergic system is in need of further exploration.

In the periphery, alpha-2 adrenergic agonists inhibit release of norepinephrine presynaptically, which might make one consider that they would have a similar effect as the postsynaptic beta-adrenergic antagonists. However, alpha-2 agonists have distinct cognitive effects. High-dose clonidine, an alpha-2 agonist, has been shown to improve immediate spatial memory in aged monkeys (Arnsten et al., 1988; Arnsten & Leslie, 1991), an effect also found in younger monkeys (Franowicz & Arnsten, 1999), believed to be mediated by action at the prefrontal cortex (Li et al., 1999). Lower doses of clonidine, those that are typically utilized clinically in humans, demonstrate varying results at varying doses, including impaired visual working memory, impulsive responses on planning tasks, and varying effects on spatial working memory (Coull et al., 1995; Jäkälä et al., 1999). Pharmacological stimulation of postsynaptic alpha-2A subtype of adrenoreceptors decreases noise and results in beneficial effects for attention deficit disorder patients (Brennan & Arnsten, 2008). However, alpha-2 agonists do not appear to have the effect on creative verbal problem solving that beta-adrenergic antagonists have (Choi et al., 2006).

Less is known about the specific cognitive effects of beta-1 and beta-2 adrenergic receptors. However, in one animal study, endogenous beta-1 selective activation impaired working memory (Ramos et al., 2005). A subsequent study demonstrated that beta-2 selective agonists enhance working memory in aging animals (Ramos et al., 2008), suggesting opposing effects between beta-1 and beta-2 receptors on working memory, and explaining the lack of effect of the nonspecific beta-antagonist propranolol on working memory in previous research (Arnsten & Goldman-Rakic, 1985; Li & Mei, 1994). Further research will be necessary to better understand the specific cognitive effects due to action at selective subtypes of beta (beta-1 and beta-2) receptors.

Other Systems

Neurons in the nucleus basalis, medial septal nucleus, and the diagonal band of Broca in the basal forebrain are the main sources of cholinergic projection throughout the neocortex and hippocampus (Selden et al., 1998). The cholinergic system is another neurotransmitter system involved in modulating the signal-to-noise ratio within the cortex by suppressing background intrinsic cortical activity (Hasselmo & Bower, 1992), thus modulating efficiency of cortical processing of sensory or associational information (Sarter & Bruno, 1997). Acetylcholine is particularly important for attentional performance (Sarter & Bruno, 2001). Studies in rodents demonstrate that acetylcholine is critical for both top-down and bottom-up processing of stimuli, mediated by action on the prefrontal cortex (Gill et al., 2000; Newman & McGaughy, 2008). Cholinergic dysfunction has been used as a model for Alzheimer's disease (Whitehouse et al., 1982), due to the significant degeneration of the cholinergic neurons in these patients. Among the two main subtypes of acetylcholine receptors, muscarinic receptors have been clearly demonstrated to interfere with encoding of new information with less of an effect on previously stored information (Hasselmo & Wyble, 1997). Blockade of nicotinic receptors has also revealed significant effects on memory in an age-dependent manner (Newhouse et al., 1992, 1994). However, despite clear effects on signal-to-noise ratio in the cortex as well as memory effects, neither muscarinic nor nicotinic blockade resulted in effects on the type of unconstrained cognitive flexibility modulated by the noradrenergic system (Smyth & Beversdorf, in preparation).

Our understanding of the role of individual neurotransmitter systems in cognition has significantly progressed over the past twenty years. However, these systems do not act in isolation. Complex interactions occur between them, which are only beginning to be understood. For example, action at D2 dopaminergic receptors and at NMDA receptors appear to interact in their effects on set shifting (Floresco et al., 2005, 2006; Stefani & Moghaddam, 2005). Also, as described above, the dopaminergic system appears to affect intradimensional set shifting (Robbins, 2007), while the noradrenergic system, specifically by action on the alpha-1 adrenergic receptor, appears to modulate performance on extradimensional set shifting (Lapiz & Morilak, 2006; Robbins, 2007). Noradrenergic innervation of dopaminergic neurons, by action on alpha-1 adrenergic receptors, is known to directly inhibit the activity of the dopaminergic neurons (Paladini & Williams, 2004). In addition, the effects of drugs on cognition also depend on location of action when isolated brain regions are studied (Cools & Robbins, 2004). Also, the mechanism by which the regulatory neurotransmitters act is beginning to be more fully understood, with potential treatment options targeting these second messenger systems (Arnsten, 2007, 2009). These factors will all need to be accounted for in future studies of creativity.

The serotonergic system, with neurons in the dorsal raphe nucleus projecting throughout the forebrain and neocortex, has long been known for its effects on mood and other psychiatric issues. However, recent research is revealing that the serotonergic system and its interaction with other neurotransmitter systems serve important cognitive roles as well. Recent evidence suggests that the balance between the serotonergic and dopaminergic systems appears to be critical for processing of reward and punishment (Krantz et al., 2010). The firing of midbrain dopamine neurons shows a firing pattern that reflects the magnitude and probability of rewards (Roesch et al., 2007; Schultz, 2007). While tryptophan depletion enhances punishment prediction but does not affect reward prediction (Cools, Robinson, & Sahakian, 2008), serotonergic neurons appear to signal reward value (Nakamura et al., 2008). Furthermore, prefrontal serotonin depletion affects reversal learning, but not set shifting (Clarke et al., 2005). A potential role in creativity is also suggested for the serotonergic system. Performance on figural and numeric creativity tasks has been associated with polymorphisms of the tryptophan hydroxide gene TPH1 (Reuter et al., 2006), and both the number solved and the prevalence of use of insight on the compound remote associates task was positively associated with high positive mood (Subramaniam et al., 2008), also suggesting a role of the serotonergic system.

Continued understanding of the roles of neurotransmitter interactions, localized effects, and other types of neurotransmitters and neuropeptides will be needed to fully understand how cognitive processes such as creativity are carried out in the brain. This will have a high potential to result in clinical benefits for patients with a wide variety of clinical syndromes as well.

References

Alexander, J. K., Hillier, A., Smith, R. M., Tivarus, M. E., & Beversdorf, D. Q. (2007). Noradrenergic modulation of cognitive flexibility during stress. *Journal of Cognitive Neuroscience, 19*, 468–478.

Angwin, A. J., Chenery, H. J., Copland, D. A., Arnott, W. L., Murdoch, B. E., & Silburn, P. A. (2004). Dopamine and semantic activation: An investigation of masked direct and indirect priming. *Journal of the International Neuropsychological Society, 10*, 15–25.

Arnsten, A. F. T. (2007). Catecholamine and second messenger influences on prefrontal cortical networks of "representational knowledge": A rational bridge between genetics and the symptoms of mental illness. *Cerebral Cortex, 17*, i6–i15.

Arnsten, A. F. T. (2009). Ameliorating prefrontal cortical dysfunction in mental illness: Inhibition of phosphotidyl inositol-protein kinase C signaling. *Psychopharmacology, 202,* 445–455.

Arnsten, A. F., & Goldman-Rakic, P. S. (1984). Selective prefrontal cortical projections to the region of the locus coeruleus and raphe nuclei in the rhesus monkey. *Brain Research, 306,* 9–18.

Arnsten, A. F., & Goldman-Rakic, P. S. (1985). Alpha-2 adrenergic mechanisms in prefrontal cortex associated with cognitive decline in aged non-human primates. *Science, 230,* 1273–1276.

Arnsten, A. F. T., Cai, J. X., & Goldman-Rakic, P. S. (1988). The alpha-2 adrenergic agonist guanfacine improves memory in aged monkeys without sedative or hypotensive side effects: evidence for alpha-2 receptor subtypes. *Journal of Neuroscience, 8,* 4287–4298.

Arnsten, A. F., Cai, J. X., Murphy, B. L., & Goldman-Rakic, P. S. (1994). Dopamine D1 receptor mechanisms in the cognitive performance of young adult and aged monkeys. *Psychopharmacology, 116,* 143–151.

Arnsten, A. F. T., & Leslie, F. M. (1991). Behavioral and receptor binding analysis of the alpha-2 adrenergic agonist, 5-bromo-6 (2-imidazoline-2-yl amino) quinoxaline (uk-14304): evidence for cognitive enhancement at an alpha-2-adrenoreceptor subtype. *Neuropharmacology, 30,* 1279–1289.

Aston-Jones, G., & Cohen, J. D. (2005). An integrative theory of locus coeruleus-norpeinephrine function: adaptive gain and optimal performance. *Annual Review of Neuroscience, 28,* 403–450.

Aston-Jones, G., Rajkowski, J., & Cohen, J. (1999). Role of locus coeruleus in attention and behavioral flexibility. *Biological Psychiatry, 46,* 1309–1320.

Barnes, C. A., & Pompeiano, M. (1991). Neurobiology of the locus coeruleus. *Progress in Brain Research, 88,* 307–321.

Belmonte, M. K., Allen, G., Beckel-Mitchener, A., Boulanger, L. M., Carper, R. A., & Webb, S. J. (2004). Autism and abnormal development of brain connectivity. *Journal of Neuroscience, 24,* 9228–9231.

Beversdorf, D. Q., Carpenter, A. L., Miller, R. F., Cios, J. S., & Hillier, A. (2008). Effect of propranolol on verbal problem solving in autism spectrum disorder. *Neurocase, 14,* 378–383.

Beversdorf, D. Q., Hughes, J. H., Steinberg, B. A., Lewis, L. D., & Heilman, K. M. (1999). Noradrenergic modulation of cognitive flexibility in problem solving. *Neuroreport, 10,* 2763–2767.

Beversdorf, D. Q., Narayanan, A., Hillier, A., & Hughes, J. D. (2007a). Network model of decreased context utilization in autism spectrum disorder. *Journal of Autism and Developmental Disorders, 37,* 1040–1048.

Beversdorf, D. Q., Ratcliffe, N. R., Rhodes, C. H., & Reeves, A. G. (1997). Pure alexia: Clinical-pathologic evidence for a lateralized visual language association cortex. *Clinical Neuropathology, 16*, 328–331.

Beversdorf, D. Q., Saklayen, S., Higgins, K. F., Bodner, K. E., Kanne, S. M., & Christ, S. E. (2011). Effect of propranolol on word fluency in autism. *Cognitive and Behavioral Neurology, 24*, 11–17.

Beversdorf, D. Q., Sharma, U. K., Phillips, N. N., Notestine, M. A., Slivka, A. P., Friedman, N. M., et al. (2007b). Effect of propranolol on naming in chronic Broca's aphasia with anomia. *Neurocase, 13*, 256–259.

Beversdorf, D. Q., White, D. M., Cheever, D. C., Hughes, J. D., & Bornstein, R. A. (2002). Central beta-adrenergic blockers modulation of cognitive flexibility. *Neuroreport, 13*, 2505–2507.

Bowden, E. M., & Jung-Beeman, M. (2003). Normative data for 144 compound remote associate problems. *Behavior Research Methods, Instruments, and Computers, 35*, 634–639.

Brennan, A. R., & Arnsten, A. F. T. (2008). Neuronal mechanisms underlying attention deficit hyperactivity disorder: The influence of arousal on prefrontal cortical function. *Annals of the New York Academy of Sciences, 1129*, 236–245.

Cahill, L., Prins, B., Weber, M., & McGaugh, J. L. (1994). β-Adrenergic activation and memory for emotional events. *Nature, 371*, 702–704.

Campbell, H. L., Tivarus, M. E., Hillier, A., & Beversdorf, D. Q. (2008). Increased task difficulty results in greater impact of noradrenergic modulation of cognitive flexibility. *Pharmacology, Biochemistry, and Behavior, 88*, 222–229.

Chamberlain, S. R., Müller, U., Blackwell, A. D., Clark, L., Robbins, T. W., & Sahakian, B. (2006a). Neurochemical modulation of response inhibition and probabilistic learning in humans. *Science, 311*, 861–863.

Chamberlain, S. R., Müller, U., Blackwell, A. D., Robbins, T. W., & Sahakian, B. (2006b). Noradrenergic modulation of working memory and emotional memory in humans. *Psychopharmacology, 188*, 397–407.

Chatterjee, A., Hamilton, R. H., & Amorapanth, P. X. (2006). Art produced by a patient with Parkinson's disease. *Behavioural Neurology, 17*, 105–108.

Chermahini, S. A., & Hommel, B. (2010). The (b)link between creativity and dopamine: Spontaneous eye blink rates predict and dissociated divergent and convergent thinking. *Cognition, 115*, 458–465.

Choi, Y., Novak, J., Hillier, A., Votolato, N. A., & Beversdorf, D. Q. (2006). The effect of α-2 adrenergic agonists on memory and cognitive flexibility. *Cognitive and Behavioral Neurology, 19*, 204–207.

Cios, J. S., Miller, R. F., Hillier, A., Tivarus, M. E., & Beversdorf, D. Q. (2009). Lack of noradrenergic modulation of indirect semantic priming. *Behavioural Neurology, 21*, 137–143.

Clarke, H. F., Walker, S. C., Crofts, H. S., Dalley, J. W., Robbins, T. W., & Roberts, A. C. (2005). Prefrontal serotonin depletion affects reversal learning but not attentional set shifting. *Journal of Neuroscience, 25*, 532–538.

Cohen, J. D., Braver, T. S., & Brown, J. W. (2002). Computational perspectives in dopamine function in prefrontal cortex. *Current Opinion in Neurobiology, 12*, 223–229.

Cools, R. (2006). Dopaminergic modulation of cognitive function-implications for L-DOPA treatment in Parkinson's disease. *Neuroscience and Biobehavioral Reviews, 30*, 1–23.

Cools, R., Gibbs, S. E., Miyakawa, A., Jagust, W., & D'Esposito, M. (2008). Working memory capacity predicts dopamine synthesis capacity in the human striatum. *Journal of Neuroscience, 28*, 1208–1212.

Cools, R., & Robbins, R. W. (2004). Chemistry of the adaptive mind. *Philosophical Transactions of the Royal Society of London, Series A: Mathematical, Physical, and Engineering Science, 362*, 2871–2888.

Cools, R., Robinson, O. J., & Sahakian, B. (2008). Acute tryptophan depletion in healthy volunteers enhances punishment prediction but does not affect reward prediction. *Neuropsychopharmacology, 33*, 2291–2299.

Copland, D. A., McMahon, K. L., Silburn, P. A., & de Zubicaray, G. I. (2009). Dopaminergic neruomodulation of semantic priming: A 4T fMRI study with levodopa. *Cerebral Cortex, 19*, 2651–2658.

Coull, J. T., Frith, C. D., Dolan, R. J., Frackowiak, R. S. J., & Grasby, P. M. (1997). The neural correlates of the noradrenergic modulation of human attention, arousal, and learning. *European Journal of Neuroscience, 9*, 589–598.

Coull, J. T., Jones, M. E. P., Egan, T. D., Frith, C. D., & Maze, M. (2004). Attentional effects of noradrenaline vary with arousal level: Selective activation of thalamic pulvinar in humans. *NeuroImage, 22*, 315–322.

Coull, J. T., Middleton, H. C., Robbins, T. W., & Sahakian, B. J. (1995). Contrasting effects of clonidine and diazepam on tests of working memory and planning. *Psychopharmacology, 120*, 311–321.

Curran, L. K., Newschaffer, C. J., Lee, L., Crawford, S. O., Johnston, M. V., & Zimmerman, A. W. (2007). Behaviors associated with fever in children with autism spectrum disorders. *Pediatrics, 120*, e1386–e1392.

de Manzano, Ö., Cervenka, S., Karabanov, A., Farde, L., & Ullén, F. (2010). Thinking outside a less intact box: thalamic dopamine D2 receptor densities are negatively related to psychometric creativity in healthy individuals. *PLoS ONE, 5*(5), e10670. doi:10.1371/journal.pone.0010670.

Dodd, M. L., Klos, K. J., Bower, J. H., Geda, Y. E., Josephs, K. A., & Ahlskog, J. E. (2005). Pathological gambling caused by drugs used to treat Parkinson disease. *Archives of Neurology, 62,* 1377–1381.

Duncan, J., Burgess, P., & Emslie, H. (1995). Fluid intelligence after frontal lobe lesions. *Neuropsychologia, 33,* 261–268.

Ehlers, A., Hackmann, A., & Michael, T. (2004). Intrusive re-experiencing in post-traumatic stress disorder: Phenomenology, theory, and therapy. *Memory, 12,* 403–415.

Ehlers, A., Hackmann, A., Steil, R., Clohessy, S., Wenninger, K., & Winter, H. (2002). The nature of intrusive memories after trauma: The warning signal hypothesis. *Behaviour Research and Therapy, 40,* 995–1002.

Eslinger, P. J., & Grattan, L. M. (1993). Frontal lobe and frontal-striatal substrates for different forms of human cognitive flexibility. *Neuropsychologia, 31,* 17–28.

Faigel, H. C. (1991). The effect of beta blockade on stress-induced cognitive dysfunction in adolescents. *Clinical Pediatrics, 30,* 441–445.

Farah, M. J., Haimm, C., Sankoorikal, G., Smith, M. E., & Chatterjee, A. (2009). When we enhance cognition with Adderall, do we sacrifice creativity? A preliminary study. *Psychopharmacology, 202,* 541–547.

Floresco, S. B., Ghods-Sharifi, S., Vexelman, C., & Magyar, O. (2006). Dissociable roles for the nucleus accumbens core and shell in regulating set shifting. *Journal of Neuroscience, 26,* 2449–2457.

Floresco, S. B., Magyar, O., Ghods-Sharifi, S., Vexelman, C., & Tse, M. T. L. (2005). Multiple dopamine receptor subtypes in the medial prefrontal cortex of the rat regulate set-shifting. *Neuropsychopharmacology, 31,* 297–309.

Foster, D. J., Good, D. C., Fowlkes, A., & Sawaki, L. (2006). Atomoxetine enhances a short-term model of plasticity in humans. *Archives of Physical Medicine and Rehabilitation, 87,* 216–221.

Franowicz, J. S., & Arnsten, A. F. T. (1999). Treatment with the noradrenergic alpha-2 agonist clonidine, but not diazepam, improves spatial working memory in normal rhesus monkeys. *Neuropsychopharmacology, 21,* 611–621.

Gallagher, D. A., O'Sullivan, S. S., Evans, A. H., Lees, A. L., & Schrag, A. (2007). Pathological gambling in Parkinson's disease: risk factors and differences from dopaminergic dysregulation: An analysis of published case series. *Movement Disorders, 22,* 1757–1763.

Ghacibeh, G. A., Shenker, J. I., Shenal, B., Uthman, B. M., & Heilman, K. M. (2006). Effect of vagus nerve stimulation on creativity and cognitive flexibility. *Epilepsy & Behavior, 8,* 720–725.

Gibbs, S. E., & D'Esposito, M. (2005). Individual capacity differences predict working memory performance and prefrontal activity following dopamine receptor stimulation. *Cognitive, Affective & Behavioral Neuroscience, 5,* 212–221.

Gill, T. M., Sarter, M., & Givens, B. (2000). Sustained visual attention performance-associated prefrontal neuronal activity evidence for cholinergic modulation. *Journal of Neuroscience, 20,* 4745–4757.

Hall, H., Sedvall, G., Magnusson, O., Kopp, J., Halldin, C., & Farde, L. (1994). Distribution of D1- and D2-dopamine receptors, and dopamine and its metabolites in the human brain. *Neuropsychopharmacology, 11,* 245–256.

Hasselmo, M. E., & Bower, J. M. (1992). Cholinergic suppression specific to intrinsic not afferent fiber synapses in rat piriform (olfactory) cortex. *Trends in Neurosciences, 67,* 1222–1229.

Hasselmo, M. E., Linster, C., Patil, M., Ma, D., & Cecik, M. (1997). Noradrenergic suppression of synaptic transmission may influence cortical signal-to-noise ratio. *Journal of Neurophysiology, 77,* 3326–3339.

Hasselmo, M. E., & Wyble, B. P. (1997). Simulation of the effects of scopolamine on free recall and recognition in a network model of the hippocampus. *Behavioural Brain Research, 89,* 1–34.

Heaton, R. K. (1981). *Wisconsin Card Sort Test manual.* Odessa, FL: Psychological Assessment Resources.

Heilman, K. M., Nadeau, S. E., & Beversdorf, D. Q. (2003). Creative innovation: Possible brain mechanisms. *Neurocase, 9,* 369–379.

Heimer, L. (1995). *The human brain and spinal cord* (2nd Ed.). New York: Springer-Verlag.

Jäkälä, P., Riekkinen, M., Sirviö, J., Koivisto, E., Kejonen, K., Vanhanen, M., & Riekkinen, P., Jr. (1999). Guanfacine, but not clonidine, improves planning and working memory performance in humans. *Neuropsychopharmacology, 20,* 460–470.

Just, M. A., Cherkassky, V. L., Keller, T. A., Kana, R. K., & Minshew, N. J. (2007). Functional and anatomical cortical underconnectivity in autism: Evidence from an fMRI study of an executive function task and corpus callosum morphometry. *Cerebral Cortex, 17,* 951–961.

Just, M. A., Cherkassky, V. L., Keller, T. A., & Minshew, N. J. (2004). Cortical activation and synchronization during sentence comprehension in high-functioning autism: Evidence of underconnectivity. *Brain, 127,* 1811–1821.

Karnath, H. O., & Wallesch, C. W. (1992). Inflexibility of mental planning: A characteristic disorder with prefrontal lobe lesions. *Neuropsychologia, 30,* 1011–1016.

Kelley, B. J., Yeager, K. R., Pepper, T. H., & Beversdorf, D. Q. (2005). Cognitive impairment in acute cocaine withdrawal. *Cognitive and Behavioral Neurology, 18,* 108–112.

Kelley, B. J., Yeager, K. R., Pepper, T. H., Bornstein, R. A., & Beversdorf, D. Q. (2007). The effect of propranolol on cognitive flexibility and memory in acute cocaine withdrawal. *Neurocase, 13,* 320–327.

Kim, N., Goel, P. K., Tivarus, M., Hillier, A., & Beversdorf, D. Q. (2010). Independent component analysis of the effect of L-dopa on fMRI of language processing. *PLoS ONE, 5*(8), e11933. doi:10.1371/journal.pone.0011933.

Kimberg, D. Y., D'Esposito, M., & Farah, M. J. (1997). Effects of bromocriptine on human subjects depend on working memory capacity. *Neuroreport, 8,* 3581–3585.

Kirschbaum, C., Pirke, K. M., & Hellhammer, D. H. (1993). The "Trier Social Stress Test"—a tool for investigating psychobiological stress responses in a laboratory setting. *Neuropsychobiology, 28,* 76–81.

Kischka, U., Kammer, T. H., Maier, S., Weisbord, M., Thimm, M., & Spitzer, M. (1996). Dopaminergic modulation of semantic network activation. *Neuropsychologia, 34,* 1107–1113.

Krantz, G. S., Kasper, S., & Lanzenberger, R. (2010). Reward and the serotonergic system. *Neuroscience, 166,* 1023–1035.

Kulisevsky, J., Pagonabarraga, J., & Martinez-Corral, M. (2009). Changes in artistic style and behaviour in Parkinson's disease: Dopamine and creativity. *Journal of Neurology, 256,* 816–819.

Kvetnansky, R., Pacak, K., Sabban, E. L., Kopin, I. J., & Goldstein, D. S. (1998). Stressor specificity of peripheral catecholaminergic activation. *Advances in Pharmacology, 42,* 556–560.

Lader, M. (1988). Beta-adrenergic antagonists in neuropsychiatry: An update. *Journal of Clinical Psychiatry, 49,* 213–223.

Lapiz, M. D. S., & Morilak, D. A. (2006). Noradrenergic modulation of cognitive function in rat medial prefrontal cortex as measured by attentional set shifting capability. *Neuroscience, 137,* 1039–1049.

Laverdue, B., & Boulenger, J. P. (1991). Medications beta-bloquantes et anxiete. Un interet therapeutique certain. [Beta-blocking drugs and anxiety. A proven therapeutic value.] *L'Encéphale, 17,* 481–492.

Li, B. M., Mao, Z. M., Wang, M., & Mei, Z. T. (1999). Alpha-2 adrenergic modulation of prefrontal cortical neuronal activity related to spatial working memory in monkeys. *Neuropsychopharmacology, 21,* 601–610.

Li, B.-M., & Mei, Z.-T. (1994). Delayed response deficit induced by local injection of the alpha-2 adrenergic antagonist yohimbine into the dosolateral prefrontal cortex in young adult monkeys. *Behavioral and Neural Biology, 62,* 134–139.

Lidow, M., Goldman-Rakic, P., Gallager, D., & Rakic, P. (1991). Distribution of dopaminergic receptors in the primate cerebral cortex: Quantitative autoradiographic analysis using (H3) raclopide, (H3) spiperone and (H3) SCH23390. *Neuroscience, 40,* 657–671.

Lipnicki, D. M., & Byrne, D. G. (2005). Thinking on your back: Solving anagrams faster when supine than when standing. *Cognitive Brain Research, 24,* 719–722.

Martchek, M., Thevarkunnel, S., Bauman, M., Blatt, G., & Kemper, T. (2006). Lack of evidence of neuropathology in the locus coeruleus in autism. *Acta Neuropathologica, 111,* 497–499.

Martindale, C., & Greenough, J. (1973). The differential effect of increased arousal on creative and intellectual performance. *Journal of Genetic Psychology, 123,* 329–335.

Mehler, M. F., & Purpura, D. P. (2009). Autism, fever, epigenetics, and the locus coeruleus. *Brain Research Reviews, 59,* 388–392.

Mehta, M. A., Manes, F. F., Magnolfi, G., Sahakian, B. J., & Robbins, T. W. (2004). Impaired set-shifting and dissociable effects on tests of spatial working memory following the dopamine D2 receptor antagonist sulpiride in human volunteers. *Psychopharmacology, 176,* 331–342.

Minderaa, R. B., Anderson, G. M., Volkmar, F. R., Akkerhuis, G. W., & Cohen, D. J. (1994). Noradrenergic and adrenergic functioning in autism. *Biological Psychiatry, 36,* 237–241.

Nakamura, K., Matsumoto, M., & Hikosaka, O. (2008). Reward-dependent modulation of neural activity in the primate dorsal raphe nucleus. *Journal of Neuroscience, 28,* 5331–5343.

Narayanan, A., White, C. A., Saklayen, S., Abduljalil, A., Schmalbrock, P., Pepper, T. H., Lander, B. N., & Beversdorf, D. Q. (2012). Functional connectivity during language processing in acute cocaine withdrawal: A pilot study. *Neurocase, 18,* 441–449.

Narayanan, A., White, C. A., Saklayen, S., Scaduto, M. J., Carpenter, A. L., Abduljalil, A., et al. (2010). Effect of propranolol on functional connectivity in autism spectrum disorder. *Brain Imaging and Behavior, 4,* 189–197.

Newhouse, P. A., Potter, A., Corwin, J., & Lenox, R. (1992). Acute nicotinic blockade produces cognitive impairment in normal humans. *Psychopharmacology, 108,* 480–484.

Newhouse, P. A., Potter, A., Corwin, J., & Lenox, R. (1994). Age-related effects of the nicotinic antagonist mecamylamine on cognition and behavior. *Neuropsychopharmacology, 10,* 93–107.

Newman, L. A., & McGaughy, J. (2008). Cholinergic deafferentation of prefrontal cortex increases sensitivity to cross-modal distractors during a sustained attention task. *Journal of Neuroscience, 28,* 2642–2650.

Paladini, C. A., & Williams, J. T. (2004). Noradrenergic inhibition of midbrain dopamine neurons. *Journal of Neuroscience, 24,* 4568–4575.

Pederzolli, A. S., Tivarus, M. E., Agrawal, P., Kostyk, S. K., Thomas, K. M., & Beversdorf, D. Q. (2008). Dopaminergic modulation of semantic priming in Parkinson disease. *Cognitive and Behavioral Neurology, 21,* 134–137.

Pessiglione, M., Czernecki, V., Pillon, B., Dubois, B., Schüpback, M., Agid, Y., et al. (2005). An effect of dopamine depletion on decision-making: The temporal coupling of deliberation and execution. *Journal of Cognitive Neuroscience, 17,* 1886–1896.

Pitman, R. K., Sanders, K. M., Zusman, R. M., Healy, A. R., Cheema, F., Lasko, N. B., et al. (2002). Pilot study of secondary prevention of posttraumatic stress disorder with propranolol. *Biological Psychiatry, 51,* 189–192.

Ramos, B. P., Colgan, L. A., Nou, E., & Arnsten, A. F. T. (2008). β2 adrenergic agonist, clenbuterol, enhances working memory performance in aging animals. *Neurobiology of Aging, 29,* 1060–1069.

Ramos, B. P., Colgan, L., Nou, E., Ovaria, S., Wilson, S. R., & Arnsten, A. F. T. (2005). The beta-1 adrenergic antagonist, betaxolol, improves working memory performance in rats and monkeys. *Biological Psychiatry, 58,* 894–900.

Ratey, J. J., Bemporad, J., Sorgi, P., Bick, P., Polakoff, S., O'Driscoll, G., et al. (1987). Brief report: open trial effects of beta-blockers on speech and social behaviors in 8 autistic adults. *Journal of Autism and Developmental Disorders, 17,* 439–446.

Reuter, M., Roth, S., Holve, K., & Hennig, J. (2006). Identification of first candidate gene for creativity: a pilot study. *Brain Research, 1069,* 190–197.

Robbins, T. W. (2007). Shifting and stopping: Fronto-striatal substrates, neurochemical modulation and clinical implications. *Philosophical Transactions of the Royal Society of London, Series B: Biological Sciences, 362,* 917–932.

Roesch, M. R., Calu, D. J., & Schoenbaum, G. (2007). Dopamine neurons encode the better option in rats between deciding between differently delayed or sized rewards. *Nature Neuroscience, 10,* 1615–1624.

Roesch-Ely, D., Weiland, S., Scheffel, H., Schwaninger, M., Hundemer, H.-P., Kolter, T., et al. (2006). Dopaminergic modulation of semantic priming in healthy volunteers. *Biological Psychiatry, 60,* 604–611.

Roozendaal, B., McReynolds, J. R., & McGaugh, J. L. (2004). The basolateral amygdala interacts with the medial prefrontal cortex in regulating glucocorticoid effects on working memory impairment. *Journal of Neuroscience, 24,* 1385–1392.

Sarter, M., & Bruno, J. P. (1997). Cognitive functions of cortical acetylcholine: Toward a unifying hypothesis. *Brain Research Reviews, 23,* 28–46.

Sarter, M., & Bruno, J. P. (2001). The cognitive neuroscience of sustained attention: Where top-down meets bottom-up. *Brain Research Reviews, 35,* 146–160.

Sawaguchi, T., & Goldman-Rakic, P. S. (1991). D1 dopamine receptors in prefrontal cortex: Involvement in working memory. *Science, 251,* 947–950.

Schultz, W. (2007). Multiple dopamine functions at different time courses. *Annual Review of Neuroscience, 30,* 259–288.

Selden, N. R., Gitelman, D. R., Salamon-Murayama, N., Parrish, T. B., & Mesulam, M.-M. (1998). Trajectories of cholinergic pathways within the cerebral hemispheres of the brain. *Brain, 121,* 2249–2257.

Silver, J. A., Hughes, J. D., Bornstein, R. A., & Beversdorf, D. Q. (2004). Effect of anxiolytics on cognitive flexibility in problem solving. *Cognitive and Behavioral Neurology, 17,* 93–97.

Smith, A., & Nutt, D. (1996). Noradrenaline and attention lapses. *Nature, 380,* 291.

Smith, R. M., & Beversdorf, D. Q. (2008). Effects of semantic relatedness on recall of stimuli preceding emotional oddballs. *Journal of the International Neuropsychological Society, 14,* 620–628.

Smyth, S. F., & Beversdorf, D. Q. (2007). Lack of dopaminergic modulation of cognitive flexibility. *Cognitive and Behavioral Neurology, 20,* 225–229.

Smyth, S. F., & Beversdorf, D. Q. (in preparation). Muscarinic and nicotinic modulation of memory but not cognitive flexibility.

Stefani, M. R., & Moghaddam, B. (2005). Systemic and prefrontal cortical NMDA receptor blockade differentially affect discrimination learning and set-shift ability in rats. *Behavioral Neuroscience, 119,* 420–428.

Stickgold, R., Hobson, J. A., Fosse, R., & Fosse, M. (2001). Sleep, learning, and dreams: Off-line memory reprocessing. *Science, 294,* 1052–1057.

Subramaniam, K., Kounios, J., Parrish, T. B., & Jung-Beeman, M. (2008). A brain mechanism for facilitation of insight by positive affect. *Journal of Cognitive Neuroscience, 21,* 415–432.

Takeuchi, H., Taki, Y., Sassa, Y., Hashizume, H., Sekiguchi, A., Fukushima, A., et al. (2010). Regional gray matter volume of dopaminergic system associate with creativity: Evidence from voxel-based morphometry. *NeuroImage, 51,* 578–585.

Tivarus, M. E., Hillier, A., Schmalbrock, P., & Beversdorf, D. Q. (2008). Functional connectivity in an fMRI study of semantic and phonological processes and the effect of L-dopa. *Brain and Language, 104,* 42–50.

Usher, M., Cohen, J. D., Servan-Schreiber, D., Rajkowski, J., & Aston-Jones, G. (1999). The role of locus coeruleus in the regulation of cognitive performance. *Science, 283,* 549–554.

Vaiva, G., Ducrocq, F., Jezequel, K., Averland, B., Lestavel, P., Brunet, A., et al. (2003). Immediate treatment with propranolol decreases posttraumatic stress disorder two months after trauma. *Biological Psychiatry, 54,* 947–949.

van Stegeren, A. H., Everaerd, W., Cahill, L., McGaugh, J. L., & Gooren, L. J. G. (1998). Memory for emotional events: differential effects of centrally versus peripherally acting β-blocking agents. *Psychopharmacology, 138,* 305–310.

Vilkki, J. (1992). Cognitive flexibility and mental programming after closed head injuries and anterior and posterior cerebral excisions. *Neuropsychologia, 30,* 807–814.

Ward, M. M., Metford, I. N., Parker, S. D., Chesney, M. A., Taylor, C. B., Keegan, D. L., et al. (1983). Epinephrine and norepinephrine responses in continuously collected human plasma to a series of stressors. *Psychosomatic Medicine, 45,* 471–486.

Whitehouse, P. J., Price, D. L., Strubble, R. G., Clark, A. W., Coyle, J. T., & DeLong, M. R. (1982). Alzheimer's disease and senile dementia—loss of neurons in the basal forebrain. *Science, 215,* 1237–1239.

Williams, G., & Goldman-Rakic, P. (1995). Modulation of memory fields by dopamine D1 receptors in prefrontal cortex. *Nature, 376,* 549–550.

Yerkes, R. M., & Dodson, J. D. (1908). The relation of strength of stimulus to rapidity of habit-formation. *Journal of Comparative Neurology and Psychology, 18,* 458–482.

9 Creativity and Psychopathology: Shared Neurocognitive Vulnerabilities

Shelley Carson

John Forbes Nash, Robert Schumann, and William Blake suffered from delusions and hallucinations (Bentley, 2001; Nasar, 1998; Ostwald, 1985). Virginia Woolf, Hart Crane, and Vincent van Gogh committed suicide during episodes of depression (Meaker, 1964). Ernest Hemingway, Henri de Toulouse-Lautrec, and William Faulkner were acknowledged alcoholics (Dardis, 1989; Frey, 1994). These are just a few of a large number of examples of creative luminaries who appear to have suffered from severe, and in all too many cases life-ending, forms of mental illness. Are these examples merely coincidental, or do they represent an actual underlying relationship between creativity and madness?

The benefits of creativity have been well documented: not only do creative innovations contribute to our survival as a species, but art, music, literature, and scientific advances add comfort and ease to daily living and enrich the human experience at both the individual and societal levels. The costs of mental illness, in terms of financial burden and human suffering, have likewise been well documented—both for the individual and for society (World Health Organization, 2001). If there is, in fact, a connection between creativity and mental illness, then understanding the nature of that connection may have broad implications in terms of the treatment of mental disorders and in the development of strategies for enhancing creative thought and accomplishment that can enrich all of humankind.

In this chapter, I will review the evidence for a creativity-psychopathology connection. I will also review several models that could account for this association. Finally, I will support a "shared neurocognitive vulnerability" model of creativity and psychopathology (Carson, 2011), in which creative ideation shares genetically influenced neurocognitive features with certain forms of mental illness. These features may manifest themselves as severe psychopathology or as creative ability, depending on the presence or absence

factors that act to protect the individual from the
...es of mental disorder.

Creativity and Mental Illness: Is There a Relationship?

Creativity is defined as an idea or product that is both novel or original and useful or adaptive in some way (Barron, 1969). Creative capacity has long been considered an advantage for humans, both at the level of the species and at the level of the individual. At the level of the species, creative ideation and behavior allow for adaptation to a changing environment and hence improved survival odds (Richards, 1999). At the level of the individual, creativity has been viewed as a facet of self-actualization (Rogers, 1961) and the expression of a fulfilled life (Maslow, 1970). It has been correlated with positive personality traits such as openness to experience and self-confidence (Feist, 1999). Creativity is also a highly valued personal trait. Besides the traditional creative fields, such as art, writing, music, and science, creativity is one of the most sought-after traits in business (Matthew, 2009), with many of the world's most prestigious business schools offering courses in creativity (Gangemi, 2006). Because creativity is so valuable, some psychologists suggest that creative accomplishment may serve as a "fitness indicator," increasing the sexual attractiveness and mating proficiency of creative individuals (Miller, 2001). In fact, studies have shown that individuals deemed creative have more sexual partners than those deemed less creative (Nettle & Clegg, 2006) and are rated as highly desirable potential mates (Buss & Barnes, 1986).

Yet despite the value of creativity at the personal and societal level, the tendency for creative individuals to suffer from what we would now call mental illness has been noted for thousands of years. Plato, for example, remarked that poets, philosophers, and dramatists had a tendency to suffer from "divine madness," one of the four types of madness cataloged in his *Phaedrus* (Plato, 360 B.C.). And Aristotle was the first to note a tendency for creative individuals to suffer from depression, as he asked in "Problem XXX" why it is that men who have become outstanding in poetry and the arts tend to be melancholic (Aristotle, 1984). The high rate of substance abuse and alcoholism in creative individuals, especially writers and poets, was noted in the early literature as well; over two thousand years ago, the Roman poet Horace wrote: "No poems can please for long or live that are written by water drinkers" (as cited in Goodwin, 1992, p. 425). The view of the mad creative genius was furthered by the moody personality and bizarre behavior of Renaissance artists like Michelangelo (Condivi, ca. 1520/1999) and the

Romantic poets (Becker, 2001). For example, Shelley suffered from hallucinations and debilitating bouts of depression, and Byron exhibited erratic mood fluctuations that ran from suicidal melancholia to extreme irritability and expansiveness (Jamison, 1993).

In the mid-twentieth century, empirical evidence for the connection between creativity and psychopathology began to emerge. In a study that examined the adopted-away offspring of mothers with and without schizophrenia, Heston (1966) found that the children of mothers with schizophrenia were more likely to hold creative jobs and have colorful lives than were the offspring of mothers without schizophrenia. Then, in a study that examined male Icelanders born between 1881 and 1910, Karlsson (1970) discovered that individuals with a psychotic relative were almost three times more likely to be registered in *Who's Who* for excellence in a creative field (scholars, novelists, poets, painters, composers, and performers) than those without a psychotic relative. He suggested that having a predisposition to schizophrenia might confer a creative advantage and concluded that "some type of mental stimulation is associated with a genetic relationship to psychotic persons" (p. 180).

These findings prompted a new generation of researchers, beginning in the late 1980s, to examine the incidence of psychopathology within the population of highly creative achievers. The results of this research support a higher risk for three categories of disorders among creative individuals: mood disorders (especially bipolar disorders), schizospectrum disorders (psychosis proneness), and substance abuse disorders. In addition, several recent studies have begun to investigate the association between creativity and the relatively new diagnosis of ADHD (Abraham, Windmann, Siefen, Daum, & Gunturkun, 2006; Cramond, 1994; White & Shah, 2006). Because the evidence for the ADHD-creativity association is not clearly established, I will limit this review to the three categories of disorders that have been more thoroughly investigated.

Creativity and Mood Disorders

The first study credited with using modern diagnostic methods to examine the creativity-psychopathology connection was conducted by noted schizophrenia researcher Nancy Andreasen (1987). Andreasen conducted diagnostic interviews with thirty writers from the prestigious Iowa Writers Workshop and their first-degree relatives and compared them to a matched control group. Based on the earlier Heston (1966) and Karlsson (1970) studies, Andreasen had expected to find a higher incidence of schizophrenia among the writers and their relatives. However, no incidents of schizophrenia were found in either the writers or the controls. Instead, she found that 80 percent of the writers suffered from a mood disorder, and that the writers were four times more likely to suffer from bipolar disorder than the controls. Andreasen also found that both mood disorders and creative interests tended to run in families, suggesting that "affective disorder may be both a 'hereditary taint' and a hereditary gift" (Andreasen, 1987, p. 1292).

Two years later, Jamison (1989) procured detailed information about mental illness, mood disorders, and creative productivity from forty-seven award-winning artists and poets from the UK. She found that an unusually high percentage of her subjects (38.3%) had been treated for mood disorder, especially bipolar disorder (6.4%). Poets (55.2%) and novelists (62.5%) had the highest rates of treatment for all types of mood disorders. She further reported that rates of creative productivity seemed to be associated with upswings in mood in both the disordered and nondisordered writers and artists. Jamison (1989) suggested that these upswings increased activation of associational networks and enhanced sensory experience, both of which are important for creativity. These findings are in line with other research that indicates that upswings in positive emotion are associated with increased expansive ideation and divergent thinking (see Ashby, Isen, & Turken, 1999, for a review).

In a subsequent study, Ludwig (1994) compared the psychiatric symptoms of fifty-nine female writers in the University of Kentucky National Women Writer's Conference with those of controls matched for age and education. The symptoms were evaluated through screenings and personal interviews using DSM III-R criteria. Rates of both depression (56%) and mania (19%) in the writers were significantly higher than in the controls.

In addition to research that examined mood disorders in living artists and writers, two studies examined mood symptoms in deceased creative luminaries, based on available biographical sources. Post (1994) studied the lives of 291 world-famous men from assorted creative professional categories and suggested psychiatric diagnoses where appropriate based on DSM III-R criteria. Using general population demographics as controls, Post found that the creative subjects in all professional categories demonstrated higher rates of mood disorder, but rates were particularly high in writers. Ludwig (1992, 1995) acquired psychiatric data from more than 1,000 deceased luminaries whose biographies had been reviewed by the *New York Times Book Review* between 1960 and 1990. He found significantly higher rates of psychopathology, including mood disorders, among persons in the

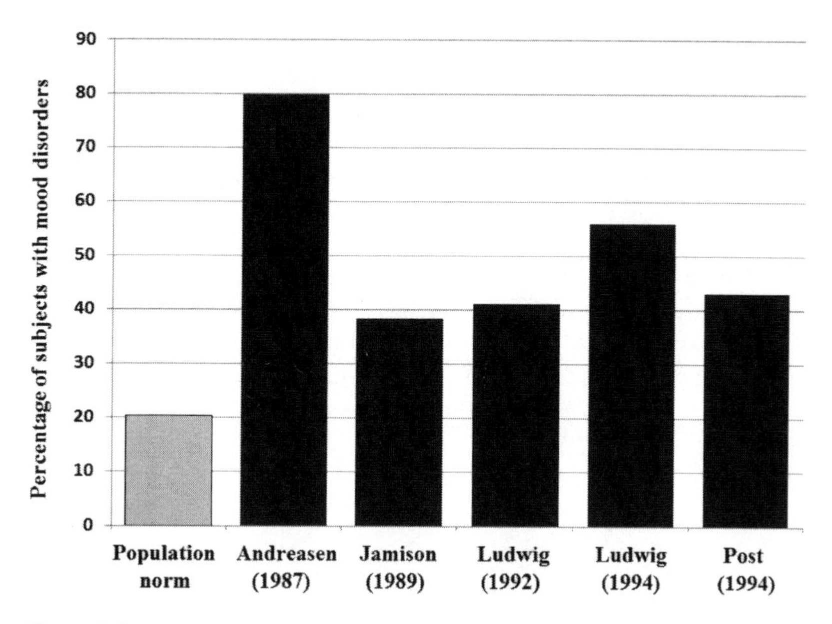

Figure 9.1

Percentage of creative individuals with mood disorder compared to population norms across studies. Population norm taken from Kessler et al. (2005).

creative arts (artists, musical composers and performers, and writers) than among those luminaries in other professions.

While these empirical studies suggest that people in creative professions (especially professions affiliated with the arts) tend to display higher rates of affective psychopathology than the norm, severe forms of affective illness may interfere with creativity. For example, writers from Andreasen's (1987) Writers Workshop study reported anecdotally that depressive episodes interfered with cognitive fluency and energy, while manic episodes led to distractibility and disorganization; both interfered with the ability to work effectively (Andreasen, 2008). Empirical evidence for the connection between creativity and the degree of affective dysfunction was provided by researchers from Harvard and Denmark (Richards, Kinney, Lunde, Benet, & Merzel, 1988). They found that subjects with cyclothymia and the first-degree relatives of subjects with manic depression had higher levels of creative accomplishment and interest than either nondisordered controls or the manic depressive subjects themselves. These authors concluded that either hereditary risk for manic depression or milder variations of bipolar pathology may enhance creativity, but that full-blown manifestations of bipolar illness may interfere with creative activity. This set of conclusions suggests an inverted "U" hypothesis relating creativity to psychopathology (Richards et al., 1988).

In general, the empirical research on creativity and mood disorders suggests (1) that creative individuals may carry a risk for affective psychopathology, especially bipolar disorder, that is greater than that of the general public (Andreasen, 1987; Jamison, 1989; Ludwig, 1992; 1994, 1995); (2) that genetic risk for (or mild forms of) affective pathology is more beneficial for creative accomplishment than more severe forms of illness (Andreasen, 2008; Richards et al., 1988); (3) that creativity and mood disorders appear to run in families (Andreasen, 1987; Jamison, 1993); and (4) that shifts in mental states associated with mood (moving toward a more positive valence) may facilitate creativity (Ashby et al., 1999; Jamison, 1989).

Creativity and Schizospectrum Disorders

Nobel Prize winner John Forbes Nash was once asked by Harvard mathematician George Mackey how he could believe that aliens from outer space had recruited him to save the world. He responded, "Because the ideas I had about supernatural beings came to me the same way that my mathematical ideas did. So I took them seriously" (quoted in Nasar, 1998, p. 11). Nash, who had been diagnosed with schizophrenia, suffered from delusions and hallucinations. His response to Mackey suggests delusional ideas may share a common element with at least some creative insights.

There is a rich literature describing psychotic and odd or eccentric behavior in creative individuals. William Blake, for example, appears to have had hallucinations since childhood; he believed that an etching technique he developed was provided to him by his dead brother, Robert, and that many of his poems and paintings were channeled to him through spirits (Galvin, 2004). The composer Robert Schumann suffered from hallucinations and delusions, and believed that Beethoven and Mendelssohn were channeling musical compositions to him from their tombs (Jensen, 2001; Lombroso, 1891/1976). Nikola Tesla, the scientist credited with discovering alternating current, suffered from auditory and visual hallucinations, including a perceived love affair with a pigeon (Pickover, 1998).

Besides anecdotal reports from the biographies of creative luminaries, studies of creative achievers at Berkeley's Institute for Personality Assessment and Research (IPAR), renowned for its groundbreaking work on creativity and personality in the 1950s and '60s, found that creative writers and creative architects had elevated scores on the Minnesota Multiphasic Personality Inventory (MMPI) scales of schizophrenia and paranoia (Barron,

1955; MacKinnon, 1962). Both creative writers and architects also reported frequent unusual perceptual occurrences and odd mystical experiences (Barron, 1969). Around the same time, a series of studies by Cattell and Drevdahl found that scientists, creative writers, and artists all scored higher on the personality measure "schizothymia" than did the presumably less creative control groups (Cattell & Drevdahl, 1955; Drevdahl & Cattell, 1958). These researchers also noted that while schizothymia was negatively correlated with self-sufficiency (a measure of good mental health) in the normal population, both schizothymia and self-sufficiency characterized their creative groups. These findings are an early indicator of the shared vulnerabilities model of creativity and psychopathology.

In the late 1970s, Robert Prentky (1979) suggested a theory that both persons with schizophrenia and highly creative persons might share a common cognitive style, namely, a style of accepting a broad bandwidth of information and processing it at a relatively shallow level of analysis, rather than focusing on a more limited volume of information and conducting detailed analysis. Evidence for this theory was provided by Dykes and McGhie (1976), who reported that creative individuals and subjects with schizophrenia tended to process auditory information in a similar manner in a dichotic listening task. Keefe and Magaro (1980) found that a group of subjects diagnosed with nonparanoid schizophrenia scored higher on two measures of a divergent thinking task than did a group of normal controls. They concluded that creative and schizophrenic individuals may share a style of thinking that has been called *loose* (Maher, 1972) or *overinclusive* (Andreasen & Powers, 1975).

However, displaying a loose cognitive style and scoring high on divergent thinking tasks is not the equivalent of accomplishing works of creativity. Researchers and creative achievers alike seem to be in agreement that meaningful creative work is difficult, if not impossible, during active episodes of schizophrenia; in the case of John Forbes Nash, his Nobel Prize–winning work was accomplished before he received the diagnosis of schizophrenia (Nasar, 1998), and the creative writers and architects in the IPAR studies had elevated scores on schizophrenia and paranoid MMPI scales rather than actual schizophrenia diagnoses. Creative individuals are more likely to exhibit subclinical traits related to psychosis rather than actual psychotic illnesses such as schizophrenia (Brod, 1987; Claridge, 1997).

Schizotypal personality—or schizotypy—is considered to be a subclinical indicator of a predisposition to psychosis; it exists on a continuum between normal experience and schizophrenia (see Claridge, 1997). A body of research has found an association between schizotypy and creativity

(Brod, 1987; Cox & Leon, 1999; Green & Williams, 1999; Poreh, Whitman, & Ross, 1994; Schuldberg, French, Stone, & Heberle, 1988). Both Brod (1987) and Prentky (1989) examined research on the connection between creativity and schizotypy/psychosis-proneness and concluded that creative persons tend to occupy a space somewhere in the midrange on the continuum between normalcy and schizophrenia. Prentky (2000–2001, p. 99) suggests that because creative individuals live "closer to the fringes of deviation, there may be a higher incidence of psychopathology among this group."

More recent studies have examined the schizotypy-creativity connection, making the distinction between *positive* and *negative* schizotypy. Positive schizotypy, or psychosis-proneness, is characterized by unusual perceptual experiences (distortions in perception, such as hearing voices in the wind) and magical thinking (fanciful ideas or paranormal beliefs, such as belief in telepathy or omens). These characteristics can be viewed as subclinical associates of hallucinations and delusions, or the *positive* signs of schizophrenia. Negative schizotypy is characterized by social anhedonia (lack of desire or pleasure in socializing with others) and cognitive disorganization (difficulties in attention, concentration, and decision making). These characteristics can be viewed as subclinical associates of the *negative* signs of schizophrenia (Mason & Claridge, 2006).

Studies by two separate research groups compared British art students to students in non-arts disciplines, finding that art students scored significantly higher than the control groups on measures of positive but not negative schizotypal traits (Burch, Pavelis, Hemsley, & Corr, 2006; O'Reilly, Dunbar, & Bentall, 2001). Nelson and Rawlings (2010) found that positive schizotypy was also associated with increased phenomenological experience of creativity in a sample of one hundred artists. Other studies indicate that positive and negative schizotypal traits may differentiate types of creative individuals. Nettle (2006) reported that poets and artists, along with psychiatric patients, had elevated levels of positive schizotypal traits, while mathematicians had higher levels of negative schizotypal traits. Along similar lines, Rawlings and Locarnini (2008) found that professional artists and musicians scored higher on measures of positive schizotypy and hypomania than biologists and mathematicians, who scored higher on many of the symptoms related to negative schizotypy. These findings are in accord with research from our Harvard lab, in which high achievers in the fields of art, music, and creative writing demonstrated significantly higher positive schizotypy scores than low achievers in those fields who were matched for IQ. High achievers in scientific fields, however, did not

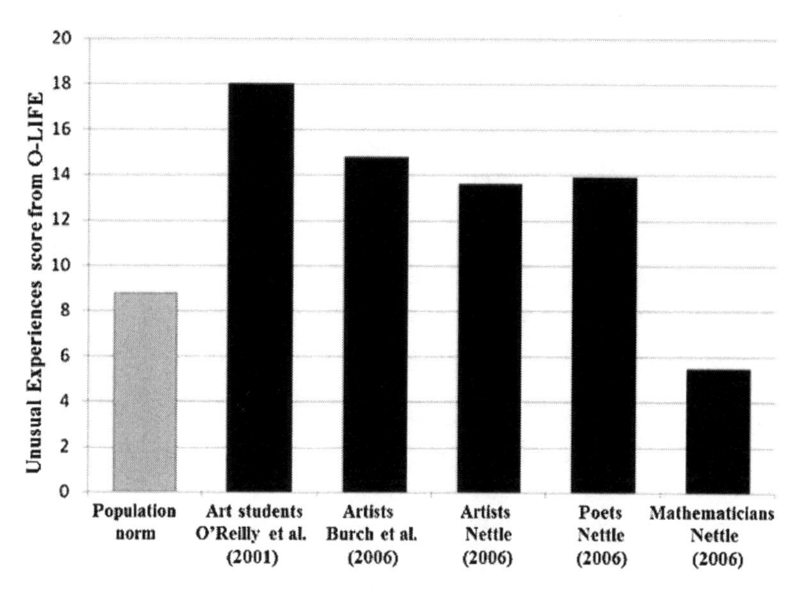

Figure 9.2

Positive schizotypy (as measured by Unusual Experiences Scale) in creative individuals compared to population norms across studies. Unusual experiences subscale of the O-LIFE and population norm taken from Mason and Claridge (2006).

demonstrate higher positive schizotypy (Carson, 2001). In sum, the results of these studies suggest that individuals in fine arts fields (artists, writers, and musicians) may have a pattern of elevated predisposition to psychosis, while scientists do not tend to show this pattern.

The "inverted U" pattern of creativity and psychopathology noted with bipolar patients (Richards et al., 1988), in which milder subclinical forms of the illness were associated with higher levels of creativity, was replicated using subjects with schizospectrum symptoms. Kinney et al. (2000–2001) found that schizospectrum traits tend to run in families. They also found that peak creativity levels were higher in subjects with schizotypal personality disorder or two schizotypy signs (such as magical ideation or illusion experiences) than in subjects with no schizotypal signs or with full-blown schizophrenia.

In general, research on schizospectrum pathology and creativity supports two conclusions: (1) there is an elevated level of schizotypy and psychosis-proneness in divergent thinkers and creative individuals (e.g., Brod, 1987; Burch et al., 2006; Schuldberg et al., 1988), and (2) milder symptom sets are more conducive to creativity than more severe forms of the schizospectrum disorders (Kinney et al., 2000–2001), as is the case with bipolar spectrum disorders.

Creativity and Alcoholism

Creative luminaries have long used alcohol and other psychoactive drugs as a method of summoning the muse. Aristophanes referred to the connection between cleverness and wine consumption (Aristophanes, 424 B.C.), suggesting that even two thousand years ago creative inspiration was associated with wine consumption. More recently, novelist William Styron described alcohol as "the magical conduit to fantasy and euphoria, and to the enhancement of the imagination" (Styron, 1990, p. 40). The implicit assumption is that alcohol consumption is related to creative output (Ludwig, 1990). Research on creativity and alcoholism does indeed indicate a greater prevalence of alcoholism among creative groups than in the general population (Andreasen, 1987; Dardis, 1989; Ludwig, 1992; Post, 1994).

Andreasen (1987) found that 30 percent of the writers in the Iowa Writers Workshop suffered from alcoholism, compared to 7 percent from the control group. Post (1994) found that 14 percent of the writers, composers, and artists in his biographical review of famous men met diagnostic criteria for alcoholism. This is about twice the rate that is found in the general public (Kessler, 2005). Ludwig (1992) also reported an elevated mean level of alcohol abuse among artists (22%), composers (21%), musical performers (40%), actors (60%), fiction writers (37%), and poets (30%), but lower than normal rates (1–2%) among natural scientists. Levels of alcoholism and alcohol abuse seem to be particularly elevated among fiction writers. Of the eight American novelists who have won the Nobel Prize, five have been alcoholics (Dardis, 1989).

As with other disorders, it appears that there is an "inverted U" association between alcoholism and creativity (Dardis, 1989; Ludwig, 1990). Biographical information from the lives of famous writers such as Hemingway, Poe, and Fitzgerald indicates that while heavy drinking and creative production went hand in hand early their careers, progressive alcoholism diminished both the quality and quantity of later creative writing (Dardis, 1989). Although creative individuals may believe that drinking inspires creativity, full-blown alcoholism appears to be detrimental to creative efforts. In a review of the effects of alcohol on the creative production of thirty-four heavy-drinking artists, writers, and composers, Ludwig (1990) found that 59 percent of the sample believed that alcohol facilitated their creativity either directly or indirectly during the early phases of their drinking. However, ultimately, 75 percent of the sample believed that

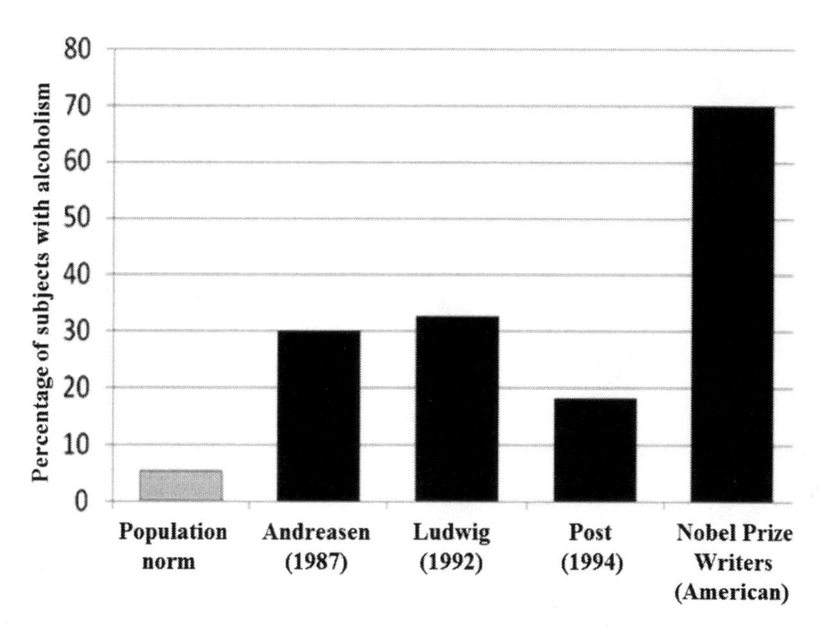

Figure 9.3
Percentage of creative individuals with alcoholism compared to population norms across studies. Population norm taken from Kessler et al. (2005).

alcohol had a direct negative effect on their work, especially in the later phases of their drinking careers.

Rates of alcoholism are elevated not only among creative individuals but among the bipolar and the schizospectrum populations as well (Kessler, Berglund, Demler, Jin, Merikangas, & Walters, 1995). A predisposition to alcoholism in creative groups as well as in groups that demonstrate forms of psychopathology related to creativity suggests an underlying shared vulnerability.

Models of the Interface between Creativity and Psychopathology

The research reviewed thus far suggests that there is both anecdotal and empirical evidence to support an elevated risk for psychopathology, especially mood disorders, schizospectrum disorders, and alcohol-related disorders, among creative individuals.

While research has persuasively indicated an association between psychopathology and creativity, the nature of the association is less clear. Several models have been proposed to account for the higher incidence of certain forms of psychopathology among highly creative individuals (e.g.,

Becker, 2001; Richards, 1990). Becker (2001), for example, has suggested that there is a social expectation that creative people will act in bizarre and unconventional ways, and that, by demonstrating symptoms of psychopathology, creative people are merely acting out the part that society has assigned them. Alternately, a "social drift" model suggests that individuals with mood disorders, alcohol problems, or psychotic tendencies do not do well in ordinary nine-to-five business routines and thus may be drawn to creative professions, such as art, writing, or music, which are unsupervised and lack rule-based expectations (Ludwig, 1995). Yet another model suggests that creative individuals may have historically been labeled as "mentally ill" in order to silence their innovative ideas and maintain the status quo.

These sociocultural models may explain some portion of the overlap between creativity and psychopathology. However, anecdotal reports (e.g., Nasar, 1998), family and adoption studies (e.g., Heston, 1966; Karlsson, 1970), as well as neuroimaging (e.g., de Manzano, Cervenka, Karabanov, Farde, & Ullén, 2010) and molecular genetic studies (e.g., Kéri, 2009) suggest that the relationship between creativity and psychopathology is not primarily sociocultural in nature. I will describe three additional models that address the creativity-psychopathology relationship at a finer-grained level of analysis.

Model 1: Creative Activity Causes (or Enhances) Mental Illness

Is there something about creativity or creative activity that could increase the risk for mental illness? Anecdotal evidence, especially from creative writers, supports such a model. For example, Hemingway suggested that creative work could be a factor in the development of depression. He wrote in his acceptance speech for the 1954 Nobel Prize in Literature, "Writing, at its best, is a lonely life ... if he is a good enough writer he must face eternity, or the lack of it, each day" (Hemingway, 1954, para. 4). Schildkraut, Hirshfeld, and Murphy (1994) studied abstract expressionist artists of the New York school, a number of whom suffered from depression and ultimately committed suicide. Schildkraut and colleagues noted, in line with Hemingway, that artists and writers must deal on a daily basis with the existential themes of the plight of man. This constant exposure to the themes of aloneness, the futility of life, and the "raw suffering of human existence" may augment depression and despair (Schildkraut et al., 1994, p. 486).

Another possible influence of creativity on psychopathology involves the stigma of being different and bucking the system, qualities that are part and parcel of creative work. Richards (1990) suggests that the rejection, mockery, and ostracism that often accompany the introduction of creative ideas can lead to feelings of alienation and subsequent psychopathology.

This model does not explain, however, why many creative individuals—whether or not they confront existential themes or face repeated rejection—do *not* exhibit signs of mental illness. In fact, a substantial body of literature suggests that artists and writers indulge in creative acts to purge themselves of *preexisting* depressive feelings (e.g., Greene, 1980; Kavaler-Adler, 1991). Creative endeavor, therefore, appears to be a self-induced therapy for, and not a cause of, psychopathology.

Model 2: Mental Illness Can Enhance Creativity

Several researchers have suggested that symptoms of certain forms of psychopathology may facilitate creativity (Hershman & Lieb, 1998; Jamison, 1993; Sass, 2000–2001). For example, Jamison (1993) proposed that the characteristics of hypomania may benefit creative idea generation. Flight of ideas may facilitate the activation of broad associational networks that is so important to creativity, and increased goal-directed activity may heighten motivation to work on creative projects. Increased energy and lack of need for sleep can provide additional resources to pursue creative activities. Meanwhile, depression, a state that has been associated with a more accurate perception of reality (see Taylor & Brown, 1988), can edit the ideas developed during hypomanic states. Depression can also serve as the subject matter for creative works, such as Emily Dickinson's famous poem "There's a Certain Slant of Light" or Tchaikovsky's "Pathétique."

Eysenck (1995) suggested that psychosis-proneness (schizotypy) may contribute to creative ideation by invoking cognitive overinclusion, a trait of schizophrenic thought. Overinclusion is a state in which the individual becomes aware of material that is typically suppressed before entering consciousness. This overinclusive state may allow associations to be made among elements that do not ordinarily appear in conscious awareness together, thus promoting unusual creative ideas. Moreover, Sass (2000–2001) has suggested that the break with reality associated with such schizotypal cognition may enhance creativity, under some circumstances, by allowing the affected individual to view situations from a totally new perspective.

Alcohol ingestion may enhance creative inspiration by reducing anxieties and inhibitions (Norlander, 1999). Such a state may allow unusual ideas to enter consciousness as normal inhibitions are relaxed. Gustafson and Norlander have provided empirical evidence for the benefit of alcohol consumption in a series of controlled studies that tested alcohol's effects on performance at various stages of the creative process (Norlander & Gustafson, 1996, 1998).

As with Model 1, this mental-illness-enhances-creativity model is not entirely satisfactory. Although certain symptoms associated with psychopathology, such as the flight of ideas, overinclusion, and reduced inhibitions, may be helpful in facilitating original or unusual ideas, the bulk of people who suffer from bipolar disorder, schizospectrum disorders, and alcohol abuse do not become unusually creative (see Richards et al., 1988).

Model 3: The Shared Vulnerability Model of Creativity and Psychopathology

A third model suggests that psychopathology and creativity may share genetic components that are expressed as either pathology or creativity depending on the presence of other moderating factors (Berenbaum & Fujita, 1994; Carson, 2011). This model could explain why highly creative individuals are at greater risk for psychopathology than the general population. It could also explain why not all highly creative individuals express psychopathology and, conversely, why not all psychosis-prone individuals express unusual creativity, as well as the findings of increased creativity in the first-degree relatives of individuals with serious psychopathology. This model could, additionally, account for the stable maintenance of a one-percent schizophrenic population within the species, when it has been demonstrated that schizophrenics reproduce fewer offspring than normal (Schlager, 1995). The place of creativity in maintaining the adaptability of the species may provide a reproductive advantage for the transmission of at least some portion of the schizophrenic genotype (Crow, 1997).

To support the model, it would be necessary to identify one or more cognitive functions or traits that act as vulnerability factors and are common to both creativity and the forms of psychopathology to which creative individuals are vulnerable. It would also be necessary to identify protective cognitive factors that are not common to both creativity and psychosis and that could, by their presence or absence, interact with the vulnerability factors to produce high levels of either creativity or psychopathology.

Evidence for the Shared Vulnerability Model of Creativity and Psychopathology

Current evidence indicates that the disorders associated with creativity, as well as creativity itself, are both heritable and polygenetic (Berrettini, 2000; Whitfield, Nightingale, O'Brien, Heath, Birley, & Martin, 1998). The polygenetic nature of creativity-related disorders suggests that a constellation of gene-related contributions are necessary for the full symptom spectrum of these disorders to be present. However, a subset of these genetic contributions may affect cognitive or emotional experience in more benign—or even highly beneficial—ways, especially if combined with a set of protective mechanisms.

Based on anecdotal reports (see Ghiselin, 1952), as well as results from neuroimaging and genetic studies, it seems reasonable to propose a model in which factors common to both creativity and psychopathology act to increase access and attention to material normally processed below the level of conscious awareness, while protective cognitive factors allow for executive monitoring and control of such enhanced access. Protective factors would thus allow creatively productive individuals to exert metacognitive control over bizarre or unusual thoughts, enabling them to take advantage of such thoughts without being overwhelmed by them (Carson, Peterson, & Higgins, 2003; Simonton, 2005). Candidates for shared vulnerability factors include reduced latent inhibition, increased novelty seeking, and neural hyperconnectivity. Candidates for protective factors include high IQ, enhanced working memory capacity, and cognitive flexibility (see figure 9.4). These

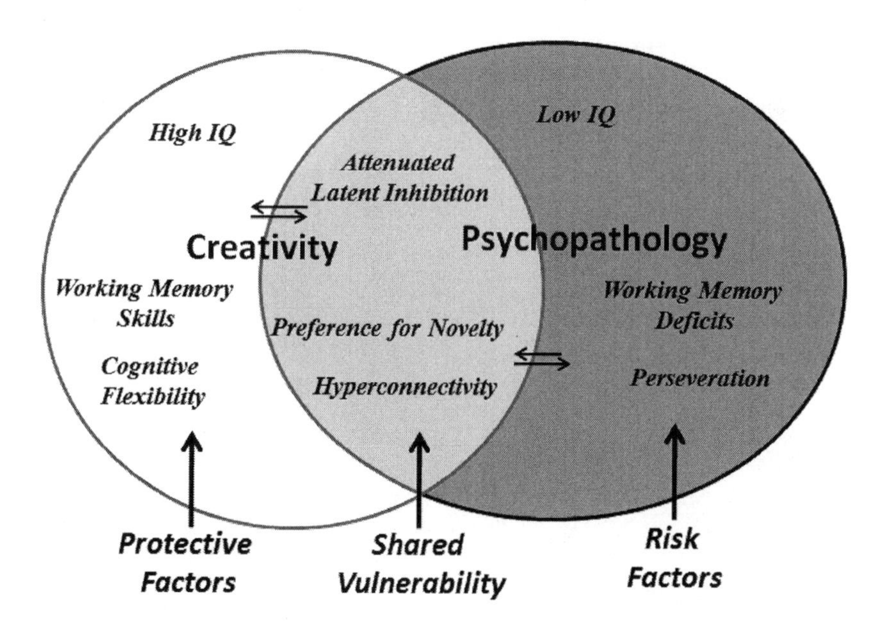

Figure 9.4
Shared vulnerability model of the creativity and psychopathology relationship. From Carson (2011). Used with permission.

are, however, part of a preliminary model that will change and expand as our knowledge of the workings of the human brain progresses.

Reduced Latent Inhibition as a Shared Neurocognitive Vulnerability Factor

Latent inhibition (LI) is the capacity to screen from conscious awareness stimuli previously experienced as irrelevant. As such, LI acts as a kind of cognitive filter. When LI is reduced, information that would typically be categorized as irrelevant is allowed into conscious awareness (Lubow & Gewirtz, 1995). Reduced LI is noted in individuals with schizophrenia and high levels of psychosis-proneness (Baruch, Hemsley, & Gray, 1988a,b; Lubow, Ingberg-Sachs, Zalstein-Orda, & Gewirtz, 1992). Reduced LI is also noted in nondisordered subjects who score high on the personality variable openness to experience, the trait often associated with creativity (Peterson & Carson, 2000), and in high-versus-low creative achievers of above average IQ (Carson, Peterson, & Higgins, 2003; Kéri, 2011). Reduced LI may enhance creativity by increasing the inventory of unfiltered stimuli available in conscious awareness, thereby improving the odds of synthesizing novel and useful combinations of stimuli (Carson et al., 2003).

Neuregulin 1 (NRG-1) has often been identified as one of the candidate genes for susceptibility to schizophrenia (Tosato, Dazzan, & Collier, 2005). Mice with a mutation of the NRG-1 gene have been shown to demonstrate reduced LI (Rimer, Barrett, Maldonado, Vock, & Gonzalez-Lima, 2005). Recently, Kéri (2009) reported that a polymorphism of the promoter region (the T/T genotype of SNP8NRG243177/rs6994992) of the NRG-1 gene that had previously been associated with psychosis was found to be prevalent in highly creative achievers with high IQs. Kéri (2009) suggested that the effect on prefrontal lobe functioning resulting from this genotype may be reduced LI, a mechanism we have already linked to both psychosis and creativity. The findings related to the NRG-1 polymorphisms support reduced LI phenomenon as a viable candidate for shared vulnerability between creativity and mental illness.

Novelty-Seeking as a Shared Neurocognitive Vulnerability Factor

Novelty-seeking is a personality trait associated with the motivation to explore novel aspects of ideas or objects. Creative individuals tend to score high in novelty-seeking and to prefer novel or complex stimuli over familiar or simple stimuli (McCrae, 1993; Reuter et al., 1995). Novelty-seeking provides internal rewards (via the dopaminergic reward system), and may arm creative individuals with intellectual curiosity and the intrinsic motivation to attend to creative work and novel ideas (Schweizer, 2006). However, novelty-seeking is also associated with alcohol abuse and addiction (Grucza et al., 2006), and with bipolar states of hypomania and mania. Novelty-seeking may be especially high in individuals with comorbid alcoholism and bipolar symptoms (Frye & Salloum, 2006). Therefore, novelty-seeking may be both an incentive for creative work and, simultaneously, a risk factor for psychopathology.

The A1+ allele of the TAQ 1A polymorphism of the DRD2 (D2 dopamine receptor) gene has been associated with novelty-seeking, schizophrenia, and addiction (Golimbet, Aksenova, Nosikov, Orlova, & Kaleda, 2003; Noble, 2000; Reuter, Schmitz, Corr, & Hennig, 2006). This same allele of the DRD2 gene was linked to creativity in a sample of German university students (Reuter, Roth, Holve, & Henning, 2006). Other genes related to dopamine functioning, including DRD4 (dopamine D4 receptor gene) and SLC6A3 (dopamine transporter gene), have also been linked to risk for both schizophrenia and bipolar disorder (Serretti & Mandelli, 2008) and to novelty-seeking (Ekelund, Lichtermann, Jarvelin, & Peltonen, 1999). Variations in novelty-seeking, as determined by the availability of dopamine in sensitive regions of the brain, may constitute another shared vulnerability factor between creative cognition and the types of psychopathology associated with creativity.

Neural Hyperconnectivity as a Shared Neurocognitive Vulnerability Factor

A third potential shared vulnerability factor, neural hyperconnectivity, is characterized by an abnormal neural linking of brain areas that are not typically functionally connected. Hyperconnectivity, perhaps caused by faulty synaptic pruning during development, has been noted in both schizophrenics and their first-degree relatives, and may be linked to the bizarre associations often reported by schizophrenics (Whitfield-Gabrieli et al., 2009). Hyperconnectivity has also been noted in neuroimaging studies of synesthesia, the tendency to make cross-modal sensory associations (Hubbard & Ramachandran, 2005). Synesthesia has a genetic component, and is seven to eight times more prevalent among highly creative individuals than in the general population (Ramachandran & Hubbard, 2001). Brang and Ramachandran (2007) have suggested that the HTR2A (serotonin transporter) gene may underlie the expression of synesthesia and may therefore be implicated in anomalous neural connectivity. This gene has also been linked to schizotypy and risk for schizophrenia, although the alleles that confer this risk have been disputed (Abdolmaleky, Faraone, Glatt, & Tsuang, 2004).

Hyperconnectivity may also be evidenced by simultaneous activation of cortical areas. Brain-imaging studies have reported more alpha synchronization, both within and across hemispheres, in the brains of highly creative versus less creative subjects during creativity tasks, suggesting unusual patterns of connectivity (Fink & Benedek, current volume; Fink et al., 2009). Ramachandran and Hubbard (2001) speculate that hyperconnectivity may form the basis of metaphorical thinking, which has been associated with creative cognition, hypomania, psychotic episodes, and drug intoxication. Unusual patterns of neural connectivity may provide one neurological mechanism for associations between disparate stimuli that are the basis of creative thought, as well as a vulnerability to mania or psychosis.

High IQ as a Protective Factor

High IQ acts as a protective factor in individuals who are vulnerable to a variety of mental disorders (Barnett, Salmond, Jones, & Sahakian, 2006). Low IQ, on the other hand, has consistently been associated with risk for schizophrenia (Woodberry, Giuliano, & Seidman, 2008). The relationship between IQ and creativity is complex; however, a body of research indicates a threshold score for IQ of around 120 is necessary but not sufficient to promote high levels of creativity (Sternberg & O'Hara, 1999). Could high IQ act as a protective factor that promotes creativity in individuals that display at least some signs of vulnerability to schizophrenia?

My colleagues and I hypothesized that if reduced LI increases the amount of stimuli available in conscious awareness, then high IQ may allow for the processing and manipulation of the additional stimuli in ways that lead to creative associations; conversely, low IQ combined with reduced LI may lead to confusion or becoming overwhelmed by the increased stimuli. In a series of studies, members of our lab found that the combination of reduced LI and high IQ predicted up to 30 percent of the variance in creative achievement scores (Carson et al., 2003) (see figure 9.5). These results suggest that when combined with reduced LI (a shared vulnerability factor), high IQ acts as a protective factor while low IQ is an additional risk factor for psychosis.

Enhanced Working Memory as a Protective Factor

Just as high IQ may allow for the advantageous processing of additional stimuli in conscious awareness due to reduced LI, enhanced working memory capacity (generally considered to be one facet of IQ) might also

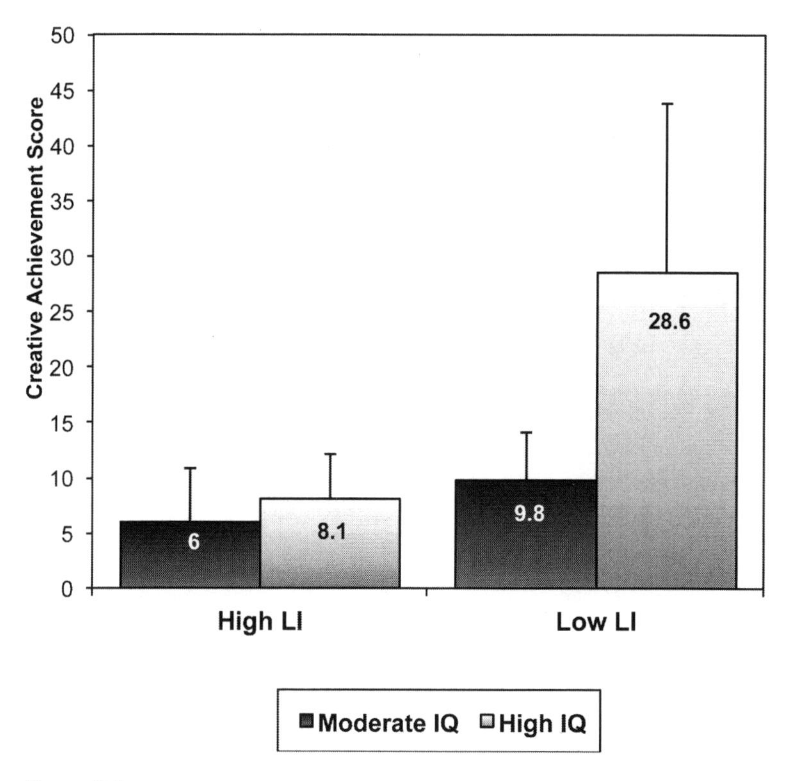

Figure 9.5
High IQ and reduced latent inhibition predict creative achievement in eminent achievers and controls. Taken from Carson et al. (2003).

confer this advantage. Support for this hypothesis emerged in a study of high-achieving student writers, composers, and artists at Harvard. LI deficits combined with high scores on a measure of working memory for abstract forms predicted over 25 percent of the variance in creative achievement scores (Carson, 2001). Working memory for abstract forms has also been shown to predict the ability to solve insight problems (a type of creativity task) in a sample of college undergraduates (DeYoung, Flanders, & Peterson, 2008). One of the most viable theories of creativity suggests that creative ideas arise from combining and rearranging bits of information that are only remotely associated with each other (Mednick, 1962). Clearly, then, the ability to hold and process a large number of bits of information simultaneously without becoming confused or overwhelmed should predispose the individual to creative rather than disordered cognition.

Cognitive Flexibility as a Protective Factor

Cognitive flexibility is the ability to disengage attention from one stimulus or concept and refocus it on another through conscious mental control. The opposite of cognitive flexibility, perseveration, is a hallmark of schizophrenia (Waford & Lewine, 2010). Cognitive flexibility may offer creative persons the ability to change perspectives and disengage from unusual thoughts or perceptions, rather than interpreting them in a psychotic manner (O'Connor, 2009).

Dietrich (2003) has suggested that creative individuals have the ability to modulate neurotransmitter systems in the brain to allow for temporary cognitive disinhibition. Evidence from several sources indicates that cognitive flexibility is dependent on dopamine availability in the prefrontal cortex (Darvas & Palmiter, 2011). Prefrontal dopamine pathways have been shown to exercise control over dopamine availability in striatal areas (Davis, Kahn, Ko, & Davidson, 1991) that, in turn, are implicated in both reduced LI and psychosis-proneness (Lubow & Gewirtz, 1995; Woodward et al., 2011). It is possible that creative individuals are able to use cognitive flexibility to promote alternating states of reduced LI and conscious executive control of attention. This may allow unusual and creative ideas to slip into consciousness awareness, while also allowing for the deliberate and rational evaluation of these ideas.

Conclusions

Despite the desirability and adaptability of human creativity, research indicates that creative individuals are at greater risk for certain forms of psychopathology than are members of the general public. Several models have been presented to account for the creativity-psychopathology relationship; however, a model of shared neurocognitive vulnerability best accounts for the available research findings.

Creative individuals may share neurocognitive vulnerabilities that are also characteristic of certain forms of mental pathology. These mechanisms may grant access to disinhibited states of consciousness, increase attention to novelty, and promote unusual associations through anomalous neural connectivity. Cognitive strengths, such as high IQ, good working memory capacity, and cognitive flexibility, may interact with these vulnerabilities to enhance creativity and to act as protective factors against severe forms of the relevant psychopathologies.

The shared vulnerability model currently includes only factors for which there is some corroborating support from brain imaging and molecular biology studies. However, there are likely additional shared vulnerabilities and protective factors that warrant inclusion. Further, the factors considered in this current model focus only on neurocognitive mechanisms. Yet recent research has emphasized the interaction of neurocognitive risk factors and environmental factors. For instance, Kéri (2011) recently demonstrated that social networks (a well-known protective factor for many mental disorders) can interact with neurocognitive mechanisms (such as reduced LI) to predict creative achievement. Future research will extend the shared vulnerability model to include the interactions of neurocognitive and environmental factors.

Clearly the inner demons of psychopathology and the voice of the muse share common characteristics within the brain and may occasionally call to the same individuals. As we learn more about the shared vulnerabilities of creativity and psychopathology, we may be able to use this knowledge to treat certain mental illnesses and effect outcomes that not only reduce human suffering but also promote human adaptability through creativity enhancement.

References

Abdolmaleky, H. M., Faraone, S. V., Glatt, S. J., & Tsuang, M. T. (2004). Meta-analysis of association between the T102C polymorphism of the 5HT2a receptor gene and schizophrenia. *Schizophrenia Research, 67,* 53–62.

Abraham, A., Windmann, S., Siefen, R., Daum, I., & Gunturkun, O. (2006). Creative thinking in adolescents with attention deficit hyperactivity disorder (ADHD). *Child Neuropsychology, 12,* 111–123.

Andreasen, N. (1987). Creativity and mental illness: Prevalence rates in writers and their first-degree relatives. *American Journal of Psychiatry, 144,* 1288–1292.

Andreasen, N. (2008). Creativity and mood disorders. *Dialogues in Clinical Neuroscience, 10,* 252–255.

Andreasen, N. J., & Powers, P. S. (1975). Creativity and psychosis. *Archives of General Psychiatry, 32,* 70–73.

Aristophanes. (424 B.C.). The knights. Retrieved from http://classics.mit.edu/Aristophanes/knights.pl.txt.

Aristotle. (1984). Problems. In J. Barnes (Ed.), *The complete works of Aristotle* (Vol. 2, pp. 1319–1527). Princeton, NJ: Princeton University Press.

Ashby, F. G., Isen, A. M., & Turken, A. U. (1999). A neuropsychological theory of positive affect and its influence on cognition. *Psychological Review, 106,* 529–550.

Barnett, G. H., Salmond, G. H., Jones, P. B., & Sahakian, B. J. (2006). Cognitive reserve in neuropsychiatry. *Psychological Medicine, 36,* 1053–1064.

Barron, F. (1955). The disposition toward originality. *Journal of Abnormal and Social Psychology, 51,* 478–485.

Barron, F. (1969). *Creative person and creative process.* New York: Holt, Rinehart & Winston.

Baruch, I., Hemsley, D. R., & Gray, J. A. (1988a). Differential performance of acute and chronic schizophrenics in a latent inhibition task. *Journal of Nervous and Mental Disease, 176,* 598–606.

Baruch, I., Hemsley, D. R., & Gray, J. A. (1988b). Latent inhibition and "psychotic proneness" in normal subjects. *Personality and Individual Differences, 9,* 777–783.

Becker, G. (2001). The association of creativity and psychopathology: Its cultural-historical origins. *Creativity Research Journal, 13,* 45–53.

Bentley, G. E., Jr. (2001). *The stranger from paradise: A biography of William Blake.* New Haven, CT: Yale University Press.

Berenbaum, H., & Fujita, F. (1994). Schizophrenia and personality: Exploring the boundaries and connections between vulnerability and outcome. *Journal of Abnormal Psychology, 103,* 148–158.

Berrettini, W. H. (2000). Susceptibility loci for bipolar disorder: Overlap with inherited vulnerability to schizophrenia. *Biological Psychiatry, 47,* 245–251.

Brang, D., & Ramachandran, V. S. (2007). Psychopharmacology of synesthesia: The role of serotonin S2a receptor activation. *Medical Hypotheses, 70,* 903–904.

Brod, J. H. (1987). Creativity and schizotypy. In G. Claridge (Ed.), *Schizotypy: Implications for illness and health* (pp. 274–298). Oxford: Oxford University Press.

Burch, G. St. J., Pavelis, C., Hemsley, D. R., & Corr, P. J. (2006). Schizotypy and creativity in visual artists. *British Journal of Psychology, 97,* 177–190.

Buss, D., & Barnes, M. (1986). Preferences in human mate selection. *Journal of Personality and Social Psychology, 50,* 559–570.

Carson, S. H. (2011). Creativity and psychopathology: A genetic shared-vulnerability model. *Canadian Journal of Psychiatry, 56,* 144–153.

Carson, S.H. (2001). *Demons and muses: An exploration of cognitive features and vulnerability to psychosis in creative individuals.* Retrieved from Dissertations and Theses database. Harvard University. (AAT3011334).

Carson, S. H., Peterson, J. B., & Higgins, D. M. (2003). Decreased latent inhibition is associated with increased creative achievement in high-functioning individuals. *Journal of Personality and Social Psychology, 85,* 499–506.

Cattell, R. B., & Drevdahl, J. E. (1955). A comparison of the personality profile (16PF) of eminent researchers with that of eminent teachers and administrators, and of the general populations. *British Journal of Psychology, 46,* 248–261.

Claridge, G. (Ed.). (1997). *Schizotypy: Implications for illness and health.* New York: Oxford University Press.

Condivi, A. (ca. 1520/1999). *The life of Michelangelo.* University Park, PA: Pennsylvania State University Press.

Cox, A. J., & Leon, J. L. (1999). Negative schizotypal traits in the relation of creativity to psychopathology. *Creativity Research Journal, 12,* 25–36.

Cramond, B. (1994). The relationship between ADHD and creativity. Paper presented at the annual meeting of the American Educational Research Association, April 2–8, 1994: New Orleans.

Crow, T. J. (1997). Is schizophrenia the price that Homo sapiens pays for language? *Schizophrenia Research, 28,* 127–141.

Dardis, T. (1989). *The thirsty muse: Alcohol and the American writer.* New York: Tichnor & Fields.

Darvas, M., & Palmiter, R. D. (2011). Contributions of striatal dopamine signaling to the modulation of cognitive flexibility. *Biological Psychiatry, 69,* 704–707.

Davis, K. L., Kahn, R. S., Ko, G., & Davidson, M. (1991). Dopamine in schizophrenia: A review and reconceptualization. *American Journal of Psychiatry, 148,* 1474–1486.

de Manzano, O., Cervenka, S., Karabanov, A., Farde, L., & Ullén, F. (2010). Thinking outside a less intact box: Thalamic dopamine d2 receptor densities are negatively related to psychometric creativity in healthy individuals. *PLoS ONE, 5,* e10670.

DeYoung, C. G., Flanders, J. L., & Peterson, J. B. (2008). Cognitive abilities involved in insight problem solving: An individual differences model. *Creativity Research Journal, 20,* 278–290.

Dietrich, A. (2003). Functional neuroanatomy of altered states of consciousness: The transient hypofrontality hypothesis. *Consciousness and Cognition, 12,* 231–256.

Dykes, M., & McGhie, A. (1976). A comparative study of attentional strategies of schizophrenic and highly creative normal subjects. *British Journal of Psychiatry, 128,* 50–56.

Drevdahl, J. E., & Cattell, R. B. (1958). Personality and creativity in artists and writers. *Journal of Clinical Personality, 14,* 107–111.

Ekelund, J., Lichtermann, D., Jarvelin, M. R., & Peltonen, L. (1999). Association between novelty seeking and the type 4 dopamine receptor gene in a large Finnish cohort sample. *American Journal of Psychiatry, 156,* 1453–1455.

Eysenck, H. J. (1995). Creativity as a product of intelligence and personality. In D. H. Saklofske & M. Zeidner (Eds.), *International handbook of personality and intelligence* (pp. 231–248). New York: Plenum Press.

Feist, G. F. (1999). The influence of personality on artistic and scientific creativity. In R. J. Sternberg (Ed.), *Handbook of creativity* (pp. 273–296). Cambridge: Cambridge University Press.

Fink, A., Grabner, R. H., Benedek, M., Reishofer, G., Hauswirth, V., Fally, M., et al. (2009). The creative brain: Investigation of brain activity during creative problem solving by means of EEG and fMRI. *Human Brain Mapping, 30*, 734–748.

Frey, J. (1994). *Toulouse-Lautrec: A life*. New York: Penguin.

Frye, M. A., & Salloum, I. M. (2006). Bipolar disorder and comorbid alcoholism: Prevalence rate and treatment considerations. *Bipolar Disorders, 8*, 677–685.

Galvin, R. (2004). William Blake: Visions and verses. *Humanities, 25*, 16–20.

Gangemi, J. (2006). Creativity comes to B-School. *Business Week*, March 26, 2006. http://www.businessweek.com/print/bschools/content/mar2006/bs20060326_8436_bs001.htm?chan_bs.

Ghiselin, B. (1952). *The creative process*. Berkeley: University of California Press.

Golimbet, V. E., Aksenova, M. G., Nosikov, V. V., Orlova, V. A., & Kaleda, V. G. (2003). Analysis of the linkage of the Taq1A and Taq1B loci of the dopamine D2 receptor gene with schizophrenia in patients and their siblings. *Neuroscience and Behavioral Physiology, 33*, 223–225.

Goodwin, D. W. (1992). Alcohol as muse. *American Journal of Psychotherapy, 46*, 422–433.

Green, M. J., & Williams, L. M. (1999). Schizotypy and creativity as effects of reduced cognitive inhibition. *Personality and Individual Differences, 27*, 263–276.

Greene, G. (1980). *Ways of escape*. New York: Simon & Schuster.

Grucza, R. A., Cloninger, C. R., Bucholz, K. K., Constantino, J. N., Schuckit, M. I., Dick, D. M., et al. (2006). Novelty seeking as a moderator of familial risk for alcohol dependence. *Alcoholism, Clinical and Experimental Research, 30*, 1176–1183.

Hemingway, E. (1954). Banquet speech. Nobel Banquet on December 10, 1954. Stockholm, Sweden. Retrieved from http://nobelprize.org/nobel_prizes/literature/laureates/1954/ hemingway-speech.html.

Hershman, D. J., & Lieb, J. (1998). *Manic depression and creativity*. New York: Prometheus.

Heston, L. L. (1966). Psychiatric disorders in foster home reared children of schizophrenic mothers. *British Journal of Psychiatry, 112*, 819–825.

Hubbard, E. M., & Ramachandran, V. S. (2005). Neurocognitive mechanisms of synaesthesia. *Neuron, 48,* 509–520.

Jamison, K. (1989). Mood disorders and patterns of creativity in British writers and artists. *Psychiatry, 52,* 125–134.

Jamison, K. R. (1993). *Touched with fire.* New York: Free Press.

Jensen, E. F. (2001). *Schumann.* New York: Oxford University Press.

Karlsson, J. L. (1970). Genetic association of giftedness and creativity with schizophrenia. *Hereditas, 66,* 177–182.

Kavaler-Adler, S. (1991). Emily Dickinson and the subject of seclusion. *American Journal of Psychoanalysis, 51,* 21–38.

Keefe, J. A., & Magaro, P. A. (1980). Creativity and schizophrenia: An equivalence of cognitive processing. *Journal of Abnormal Psychology, 89,* 390–398.

Kéri, S. (2009). Genes for psychosis and creativity: A promoter polymorphism of the neuregulin 1 gene is related to creativity in people with high intellectual achievement. *Psychological Science, 20,* 1070–1073.

Kéri, S. (2011). Solitary minds and social capital: Latent inhibition, general intellectual functions and social network size predict creative achievements. *Psychology of Aesthetics, Creativity, and the Arts, 5,* 215–221.

Kessler, R. C., Berglund, P., Demler, O., Jin, R., Merikangas, K. R., & Walters, E. E. (2005). Lifetime prevalence and age-of-onset distribution of *DSM-IV* disorders in the National Comorbidity Survey replication. *Archives of General Psychiatry, 62,* 593–602.

Kinney, D. K., Richards, R., Lowing, P. A., LeBlanc, D., Zimbalist, M. E., & Harlan, P. (2000–2001). Creativity in offspring of schizophrenic and control parents: An adoption study. *Creativity Research Journal, 13,* 17–25.

Lombroso, C. [1891] (1976). *The man of genius.* London: Walter Scott.

Lubow, R. E., & Gewirtz, J. C. (1995). Latent inhibition in humans: Data, theory, and implications for schizophrenia. *Psychological Bulletin, 117,* 87–103.

Lubow, R. E., Ingberg-Sachs, Y., Zalstein-Orda, N., & Gewirtz, J. C. (1992). Latent inhibition in low and high "psychotic-prone" normal subjects. *Personality and Individual Differences, 13,* 563–572.

Ludwig, A. (1990). Alcohol input and creative output. *British Journal of Addiction, 85,* 953–963.

Ludwig, A. (1992). Creative achievement and psychopathology: Comparison among professions. *American Journal of Psychotherapy, 46,* 330–354.

Ludwig, A. (1994). Mental illness and creative activity in female writers. *American Journal of Psychiatry, 151,* 1650–1656.

Ludwig, A. (1995). *The price of greatness: Resolving the creativity and madness controversy.* New York: Guilford Press.

MacKinnon, D. W. (1962). The nature and nurture of creative talent. *American Psychologist, 17,* 484–495.

Maher, B. (1972). The language of schizophrenia: A review and interpretation. *British Journal of Psychiatry, 120,* 3–17.

Maslow, A. H. (1970). *Motivation and personality* (2nd Ed.). New York: Harper & Row.

Mason, O., & Claridge, G. (2006). The Oxford-Liverpool Inventory of Feelings and Experiences (O-LIFE): Further description and extended norms. *Schizophrenia Research, 82,* 203–211.

Matthew, C. A. (2009). Leader creativity as a predictor of leading change in organizations. *Journal of Applied Social Psychology, 39,* 1–41.

McCrae, R. R. (1993). Openness to experience as a basic dimension of personality. *Imagination, Cognition, and Personality, 13,* 39–55.

Meaker, M. J. (1964). *Sudden endings: 13 profiles in depth of famous suicides.* Garden, NY: Doubleday.

Mednick, S. (1962). The associative basis of the creative process. *Psychological Review, 69,* 220–232.

Miller, G. F. (2001). Aesthetic fitness: How sexual selection shaped artistic virtuosity as a fitness indicator and aesthetic preference as mate choice criteria. *Bulletin of Psychology and the Arts, 2,* 20–25.

Nasar, S. (1998). *A beautiful mind: The life of mathematical genius and Nobel laureate John Nash.* New York: Simon & Schuster.

Nelson, B., & Rawlings, D. (2010). Relating schizotypy and personality to the phenomenology of creativity. *Schizophrenia Bulletin, 36,* 388–399.

Nettle, D. (2006). Schizotypy and mental health amongst poets, visual artists, and mathematicians. *Journal of Research in Personality, 40,* 876–890.

Nettle, D., & Clegg, H. (2006). Schizotypy, creativity, and mating success in humans. *Proceedings of the Royal Society of London, Series B: Biological Sciences, 273,* 611–615.

Noble, E. P. (2000). Addiction and its reward process through polymorphisms of the D2 dopamine receptor gene: A review. *European Psychiatry, 15,* 79–89.

Norlander, T. (1999). Inebriation and inspiration? A review of the research on alcohol and creativity. *Journal of Creative Behavior, 33*, 22–44.

Norlander, T., & Gustafson, R. (1998). Effects of alcohol on a divergent figural fluency test during the illumination phase of the creative process. *Creativity Research Journal, 11*, 265–274.

Norlander, T., & Gustafson, R. (1996). Effects of alcohol on scientific thought during the incubation phase of the creative process. *Journal of Creative Behavior, 30*, 231–248.

O'Connor, K. (2009). Cognitive and meta-cognitive dimensions of psychoses. *Canadian Journal of Psychiatry, 54*, 152–159.

O'Reilly, T., Dunbar, R., & Bentall, R. (2001). Schizotypy and creativity: An evolutionary connection. *Personality and Individual Differences, 31*, 1067–1078.

Ostwald, P. (1985). *Schumann: The inner voices of a musical genius*. Boston: Northeastern University Press.

Peterson, J. B., & Carson, S. (2000). Latent inhibition and openness to experience in a high-achieving student population. *Personality and Individual Differences, 28*, 323–332.

Pickover, C. A. (1998). *Strange brains and genius*. New York: Plenum Press.

Plato. (360 B.C.) *Phaedrus*. MIT Internet Classics. http://classics.mit.edu/Plato/phaedrus.html.

Poreh, A. M., Whitman, D. R., & Ross, T. P. (1994). Creative thinking abilities and hemispheric asymmetry in schizotypal college students. *Current Psychology, 12*, 344–352.

Post, F. (1994). Creativity and psychopathology: A study of 291 world-famous men. *British Journal of Psychiatry, 165*, 22–34.

Prentky, R. (1979). Creativity and psychopathology: A neurocognitive perspective. In B. Maher (Ed.), *Progress in experimental personality research* (Vol. 9, pp. 1–39). New York: Academic Press.

Prentky, R. (1989). Creativity and psychopathology: Gamboling at the seat of madness. In J. A. Glover, R. R. Ronning, & C. R. Reynolds (Eds.), *Handbook of creativity* (pp. 243–270). New York: Plenum Press.

Prentky, R. (2000–2001). Mental illness and roots of genius. *Creativity Research Journal, 13*(1), 95–104.

Ramachandran, V. S., & Hubbard, E. M. (2001). Synaesthesia—a window into perception, thought, and language. *Journal of Consciousness Studies, 8*, 3–34.

Rawlings, D., & Locarnini, A. (2008). Dimensional schizotypy, autism, and unusual word associations in artists and scientists. *Journal of Research in Personality, 42,* 465–471.

Reuter, M., Panksepp, J., Schnabel, N., Kellerhoff, N., Kempel, P., & Hennig, J. (1995). Personality and biological markers of creativity. *European Journal of Personality, 19,* 83–95.

Reuter, M., Roth, S., Holve, K., & Henning, J. (2006a). Identification of first genes for creativity: A pilot study. *Brain Research, 1069,* 190–197.

Reuter, M., Schmitz, A., Corr, P., & Hennig, J. (2006b). Molecular genetics support Gray's personality theory: The interaction of COMT and DRD2 polymorphisms predicts the behavioural approach system. *International Journal of Neuropsychopharmacology, 9,* 155–166.

Richards, R. (1999). Everyday creativity. In M. A. Runco & S. R. Pritzker (Eds.), *Encyclopedia of creativity* (Vol. 1, pp. 683–687). San Diego: Academic Press.

Richards, R. (1990). Everyday creativity, eminent creativity, and health: "Afterview" for CRJ issues on creativity and health. *Creativity Research Journal, 3,* 300–326.

Richards, R., Kinney, D. K., Lunde, I., Benet, M., & Merzel, A. P. C. (1988). Creativity in manic—depressives, cyclothymes, their normal relatives, and control subjects. *Journal of Abnormal Psychology, 97,* 281–288.

Rimer, M., Barrett, D. W., Maldonado, M. A., Vock, V. M., & Gonzalez-Lima, F. (2005). Neuregulin-1 immunoglobulin-like domain mutant mice: Clozapine sensitivity and impaired latent inhibition. *Neuroreport, 16,* 271–275.

Rogers, C. (1961). *On becoming a person: A therapist's view of psychotherapy.* London: Constable.

Sass, L. A. (2000–2001). Schizophrenia, modernism, and the "creative imagination": On creativity and psychopathology. *Creativity Research Journal, 13,* 55–74.

Schildkraut, J., Hirshfeld, A., & Murphy, J. (1994). Mind and mood in modern art, II: Depressive disorders, spirituality, and early deaths in the abstract expressionist artists of the New York school. *American Journal of Psychiatry, 151,* 482–488.

Schlager, D. (1995). Evolutionary perspectives on paranoid disorder. *Psychiatric Clinics of North America, 18,* 263–279.

Schuldberg, D., French, C., Stone, B. L., & Heberle, J. (1988). Creativity and schizotypal traits: Creativity test scores and perceptual aberration, magical ideation, and impulsive nonconformity. *Journal of Nervous and Mental Disease, 176,* 648–657.

Schweizer, T. J. (2006). The psychology of novelty-seeking, creativity, and innovation: Neurocognitive aspects within a work—psychological perspective. *Creativity and Innovation Management, 15,* 164–172.

Serretti, A., & Mandelli, L. (2008). The genetics of bipolar disorder: Genome "hot regions," genes, new potential candidates, and future directions. *Molecular Psychiatry, 13,* 742–771.

Simonton, D. K. (2005). Are genius and madness related? Contemporary answers to an ancient question. *Psychiatric Times, 22,* 21–22.

Sternberg, R. J., & O'Hara, L. A. (1999). Creativity and intelligence. In R. J. Sternberg (Ed.), *Handbook of creativity* (pp. 251–272). New York: Cambridge University Press.

Styron, W. (1990). *Darkness visible: A memoir of madness.* New York: Random House.

Taylor, S. E., & Brown, J. D. (1988). Illusion and well-being—a social psychological perspective on mental-health. *Psychological Bulletin, 103,* 193–210.

Tosato, S., Dazzan, P., & Collier, D. (2005). Association between the neuregulin 1 gene and schizophrenia: A systematic review. *Schizophrenia Bulletin, 31,* 613–617.

Waford, R. N., & Lewine, R. (2010). Is perseveration uniquely characteristic of schizophrenia? *Schizophrenia Research, 118,* 128–133.

White, H. A., & Shah, P. (2006). Uninhibited imaginations: Creativity in adults with attention deficit/hyperactivity disorder. *Personality and Individual Differences, 40,* 1121–1131.

Whitfield, J. B., Nightingale, B. N., O'Brien, M. E., Heath, A. C., Birley, A. J., & Martin, N. G. (1998). Molecular biology of alcohol dependence: A complex polygenic disorder. *Clinical Chemistry and Laboratory Medicine, 36,* 633–636.

Whitfield-Gabrieli, S., Thermenos, H. W., Milanovic, S., Tsuang, M. T., Faraone, S. V., McCarley, R. W., et al. (2009). Hyperactivity and hyperconnectivity of the default network in schizophrenia and in first-degree relatives of persons with schizophrenia. *Proceedings of the National Academy of Sciences of the United States of America, 106,* 1279–1284.

Woodberry, K. A., Giuliano, A. J., & Seidman, L. J. (2008). Premorbid IQ in schizophrenia: A meta-analytic review. *American Journal of Psychiatry, 165,* 579–587.

Woodward, N. D., Cowan, R. L., Park, S., Ansari, M. S., Baldwin, R. M., Li, R., et al. (2011). Correlation of individual differences in schizotypal personality traits with amphetamine-induced dopamine release in striatal and extrastriatal brain regions. *American Journal of Psychiatry, 168,* 418–426.

World Health Organization. (2001). *The World Health Report 2001: Mental health: New understanding, new hope.* Geneva: World Health Organization.

V Neuroimaging

10 The Creative Brain: Brain Correlates Underlying the Generation of Original Ideas

Andreas Fink and Mathias Benedek

It is commonly believed that the ability to think creatively is advantageous in a variety of areas of our everyday life. Creativity—defined as the ability to produce work that is both novel (original, unique) and useful within a social context (e.g., Flaherty, 2005; Stein, 1953; Sternberg & Lubart, 1996)—appears to be crucial in culture, science, and education, as well as the economical or industrial domain. Employees are required to produce novel and innovative ideas. By the same token, pedagogues and teachers instruct their students to produce creative work or achievements. Similarly, in constructing buildings, furnishing workplaces or homes, or even in creating our outfit we continually rely on creativity-related skills.

Even though the striking role of creativity in these areas appears to be beyond dispute, our scientific understanding of this topic lags behind. In fact, creativity has (unlike other mental ability constructs such as intelligence) long been viewed as a "difficult" trait that is hardly amenable to research, and empirical studies on this topic were extremely scarce. In 1950, Guilford's seminal address to the American Psychological Association brought about a resurgence in this research field. His most influential contribution to this field was presumably in a conceptual as well as in a psychometrical sense, inasmuch as he specified several characteristics of creative people that could be measured by means of psychometric tests. According to Guilford (1950), creative people might be characterized by ideational fluency (i.e., showing a large quantity of ideas), originality (i.e., novelty or uniqueness of ideas), or the ability to think flexibly (i.e., the ability to produce different types of ideas). Stimulated by Guilford's work, many creativity measures have been developed and empirically tested, among the most prominent being the Torrance Tests of Creative Thinking (TTCT; Torrance, 1966), or the divergent production tests by Guilford (1967). The availability of creativity measures has in turn stimulated relevant research activities in several scientific disciplines and, in the meanwhile, creativity has been addressed from a variety of different perspectives. It has, for instance, been studied in the cognitive sciences (e.g., Smith et al., 1995; Ward, 2007), in pedagogy or the educational domain (e.g., Sawyer, 2006), from the perspective of social psychology (e.g., Amabile, 1983; Hennessey & Amabile, 2010), in the context of mental illness (e.g., Kaufman, 2005), and most recently in the field of neuroscience (see, e.g., Arden et al., 2010; Dietrich, 2007; Dietrich & Kanso, 2010; Fink et al., 2007; Jung et al., 2010).

This chapter attempts to show how neuroscientific approaches in this field can help us to learn more about this extremely important research topic that has been neglected for a relatively long period of time. In doing so, we will first briefly summarize recent research in this field which aims at investigating how different facets of creative cognition are manifested in our brains. Second, we will address the crucial research question as to how creative cognition can be improved effectively by training or cognitive stimulation, and whether any intervention effects are also observable at the level of the brain. We hope to demonstrate that neuroscientific research approaches in this field are a valuable tool for improving our scientific understanding of this complex but nevertheless important and fascinating mental domain.

The Neuroscientific Study of Creative Cognition

Neuroimaging techniques such as functional magnetic resonance imaging (fMRI), near infrared spectroscopy (NIRS), the measurement of the brain's glucose metabolism via positron emission tomography (PET), or the analysis of different parameters in the electroencephalogram (EEG) allow us to investigate brain activity during a broad range of different cognitive demands. Each of these neuroscientific measurement methods has its pros and cons in the particular context of the study of creative cognition. The primary advantage of fMRI lies in its high spatial accuracy, but it does not allow for the study of cognition with high temporal resolution (as opposed to EEG techniques). The observed changes in brain activity (e.g., blood-oxygen-level dependent [BOLD] response from a prestimulus reference condition to an activation interval) occur rather slowly, thereby complicating the analysis of time-related brain activity patterns during the process of creative thinking. EEG techniques, in contrast, show lower spatial resolution but allow for a fine-grained temporal analysis of brain activation that is observed in response to a particular cognitive event (e.g., immediately prior to the production of an original idea). Also, in analyzing the functional cooperation (or functional coupling, respectively) between different cortical areas EEG techniques have turned out to be a valuable tool in the study of creative cognition (see e.g., Bhattacharya & Petsche, 2005; Grabner et al., 2007; Jaušovec, 2000; Jaušovec & Jaušovec, 2000; Mölle et al., 1999; Petsche, 1996; Razumnikova, 2000; Sandkühler & Bhattacharya, 2008).

The EEG signal represents oscillations observed across a wide range of frequencies that are commonly divided into distinct frequency bands (e.g., alpha band: 8–13 Hz, beta band: 13–30 Hz). Spectral analyses of the EEG can be used to compute the band-specific frequency power for given periods of time. Additionally, task- or event-related power changes can be quantified by contrasting the power in a specified frequency band during a cognitive task with a preceding reference interval. Event-related power decreases from a reference to an activation interval are referred to as event-related desynchronization (ERD), while power increases are referred to event-related synchronization (ERS; Pfurtscheller, 1999). ERD/ERS of the alpha band has been found to be especially sensitive to cognitive task performance and higher cognitive abilities (for a review, see Klimesch, 1999). In several studies of our laboratory we investigated task- or event-related power changes in the EEG alpha band while individuals were engaged in the performance of different types of creative idea-generation tasks. The construction of the creativity tasks (or the modification of selected tasks from literature, respectively) was strictly guided by the aim to realize them appropriately in the neurophysiological laboratory. Relevant studies in this field (regardless of whether fMRI, PET, NIRS, or EEG is used) are challenged by the realization of suitable, ecologically valid experimental paradigms or procedures that try to resemble "real-life" creativity (outside the lab) to the best possible extent. At the same time, however, drawing, free-hand writing, speaking, or any other body movements and the like that could negatively influence the quality of the neurophysiological measurements have to be minimized or avoided. In this particular context one might be well advised to separate a creativity task into time intervals during which participants are required to think creatively (creative idea generation phase) and time intervals during which participants are required to (orally) respond to the given stimulus. This would enable the experimenter to investigate functional patterns of brain activity during time periods of creative thinking that are free of, or at least less prone to, artifacts caused by speaking or free-hand writing. At the same time, the oral responses that are recorded subsequent to the creative idea generation period (and transcribed for analyses by the experimenter) allow for a reliable analysis of behavioral task performance.

In following such an approach, four different types of creative idea-generation tasks were adapted and empirically tested in our lab. In the classic alternative uses test (AU), the participants' task is to name as many and original uses of a conventional, everyday object (e.g., unusual uses of a "brick" or "tin"). In the insight task (IS), participants are confronted with unusual, hypothetical situations in need of explanation (e.g., *"A light in the darkness"*). They are required to think of as many different causes, reasons, or conditions as possible that may explain the given situation. Similarly, in the utopian situation task (US), participants are instructed to put themselves in the given utopian situations and to produce as many and as original consequences as possible that would arise from this situation (e.g., *"Imagine there were a creeping plant rising up to the sky. What would await you at the end of this plant?"*). And finally, in the word ends task (WE), German suffixes are presented that have to be completed by the participants in many different ways. In all of these tasks participants were instructed to produce many different (i.e., ideas out of different categories) and original ideas (i.e., ideas no one else would think of). Brain activity during creative idea generation (i.e., during the creative idea generation period, in which participants were requested to [silently] think of possible solutions) was quantified by means of task- or event-related power changes in different EEG alpha-frequency bands.

Which Brain Correlates Are Associated with the Generation of Creative Ideas?

Behavioral analyses of the employed creative idea generation tasks revealed that they differ notably with respect to their task demands. This was evident by the finding that performance in the IS, US, and AU task (as opposed to the WE task) was more strongly correlated with the NEO FFI (Neuroticism Extraversion Openness Five Factor Inventory) factor "openness to experiences" which is seen in relation to creativity (e.g., Feist, 1998; King et al., 1996). In contrast, completing suffixes (i.e., performance of the WE task) was significantly correlated with verbal intelligence (Benedek et al., 2006; Fink et al., 2006, 2007), whereas in the IS, US, and AU tasks no correlation with verbal ability was apparent at all. Thus, the IS, US, and AU tasks seem to rely more strongly on divergent, free-associative demands, whereas the WE task rather involves convergent, intelligence-related demands.

Most interestingly, task differences were not only apparent on the behavioral level but were observed on the neurophysiological level as well.

The IS, US, and AU tasks were accompanied by relatively strong synchronization of alpha activity, whereas in the WE task, which was correlated with verbal intelligence, the lowest synchronization of alpha activity was observed. Though topographically less restricted, these increases in alpha activity were most apparent in posterior (parietal) regions of the brain. These findings suggest that the more creativity-related a task is (e.g., finding original alternate solutions as opposed to completing suffixes) the stronger is the synchronization of alpha activity (Fink et al., 2007; a discussion of the functional meaning of task or event-related alpha synchronization can be found later on in this chapter). This finding is nicely in line with other EEG studies suggesting that convergent (i.e., more intelligence-related) versus divergent (i.e., more creativity-related) modes of thinking are accompanied by different activity patterns of the brain. The study by Mölle et al. (1999), for instance, yields evidence that divergent thinking tasks (e.g., alternate uses task) evoke higher EEG complexity (i.e., estimated correlation dimension of EEG as determined by means of singular value decomposition of the EEG signal) than convergent thinking tasks strongly drawing on intelligence (e.g., mental arithmetic), which could be the result of a larger number of independently oscillating neural assemblies during the former type of thinking. Further evidence for differences in EEG patterns between convergent and divergent thinking tasks was also reported by Jauk et al. (2012) and Jaušovec and Jaušovec (2000), as well as Razumnikova (2000, 2004).

The finding that brain activity during creative cognition might be different from brain activity observed in response to other types of cognition could be also confirmed in a study by Carlsson et al. (2000) that measured regional cerebral blood flow (rCBF). The authors showed that when participants were required to name as many different uses of bricks, they displayed a higher level of prefrontal brain activation than during the performance of a more intelligence-related verbal fluency task (naming words that begin with a given letter). Furthermore, using fMRI, Goel and Vartanian (2005) report evidence that solving match problems (known as a creative problem-solving task) exhibited activation (relative to a convergent baseline condition) in the left dorsal lateral as well as in the right ventral lateral prefrontal cortex. Thus, neuroscientific studies that contrast the performance of more creativity-related tasks with the performance of more intelligence-related problem-solving tasks provide evidence that these different modes of thinking are accompanied by qualitatively different activity patterns of the brain.

Investigation of Brain Activity in Response to the Generation of More versus Less Original Ideas

Another promising approach in the neuroscientific study of creative cognition is to investigate brain activity patterns that are associated with the production of highly creative (as opposed to less creative) ideas. This exciting research question has been stimulated by Jung-Beeman et al. (2004), who investigated brain correlates underlying the subjective experience of a flash of insight, or "*aha*!" The authors had their participants work on remote associate problems (finding a compound to three given unrelated words; e.g., for "rat—blue—cottage," the solution would be "cheese") and compared brain activity during solutions that were accompanied by a subjective experience of "*aha*!" with those that were solved without the subjective experience of insight. Stimulated by Jung-Beeman et al.'s approach, Fink and Neubauer (2006) investigated how brain states during the production of highly original ideas might be differentiated from those observed during the production of less original ideas. To obtain a measure of originality of the responses given during the performance of experimental tasks, we applied an external rating procedure similar to Amabile's (1982) consensual assessment technique that has frequently been employed in this field of research. To this end, three female and three male raters were asked to evaluate the responses of the participants given during the experiment with respect to their originality on a five-point rating scale (ranging from 1, "highly original," to 5, "not original at all"). Subsequently the ratings were averaged over all raters (who displayed satisfying internal consistency in their ratings), separately for each idea, so that one originality score was available for each single response of a task. Based on the external ratings, we compiled lists of high-original versus low-original ideas within each single task and participant (by means of a median split). Analyses revealed that more original (as opposed to less original) ideas were accompanied by a stronger synchronization of alpha activity in centroparietal regions of the cortex. This finding is in agreement with Jung-Beeman et al.'s (2004) "alpha effect" observed during subjective experience of "*aha*!" (viz., an increase of parietal alpha activity in insight as compared to noninsight solutions). Moreover, this result fits into previous research reports which also found parietal brain regions being critically involved in divergent or creative cognition tasks (e.g., Bechtereva et al., 2004; Razumnikova, 2004).

In Grabner et al. (2007), we aimed at extending the Fink and Neubauer (2006) findings in two important ways. First, we assessed brain activity in relation to *self-rated* originality of ideas. Participants were asked to evaluate each idea they gave during the experiment with respect to its originality. This was realized subsequent to the performance of the creative idea-generation tasks (i.e., subsequent to the recording session). Again, a five-point rating scale ranging from 1 (highly original) to 5 (not original at all) was used, but, in contrast to Fink and Neubauer (2006), the originality scores were now based on subjective instead of external ratings. And second, as creative cognition presumably requires different functional neural networks distributed over the whole brain, we also used measures informing us about the functional cooperation between different brain areas. For this reason, we calculated functional coupling or the phase locking value (PLV) between selected pairs of electrodes. As outlined by Lachaux et al. (1999), the PLV measures the covariance of phase between two different neuroelectrical signals. Because we were interested in changes of phase synchrony in response to the generation of creative ideas, we computed event-related phase-locking values in applying the same formula as for calculating ERD/ERS.

Several findings of the Grabner et al. (2007) study appear to be noteworthy: First, similar to Fink and Neubauer (2006), creative idea generation was generally accompanied by an event-related synchronization (ERS) of alpha activity. Second, and more importantly, the obtained findings also suggest that the production of ideas that were subjectively rated as more original was reflected in a different activity pattern of the brain than the production of less original ideas. Analyses revealed that the production of more original ideas exhibited a larger right-hemispheric ERS in the lower alpha band (8–10 Hz) than the production of less original ideas, whereas in the left hemisphere no ERS differences in relation to self-rated originality of ideas were found. Interestingly, differences between subjectively more versus less original responses were observed not only with respect to ERS but also with respect to event-related phase locking. Findings suggest that more original ideas were associated with a larger event-related PLV in anterior cortices of the right hemisphere, while in the left hemisphere no significant PLV differences between more and less original responses emerged. As will be discussed in more detail below, these findings support the view of the frontal cortex as key brain region in creative cognition (for reviews, see Dietrich & Kanso, 2010; Heilman et al., 2003). Frontal cortices are believed to be involved in processes such as cognitive flexibility, attention, semantic information processing, and working memory, which may likewise play a crucial role in creativity. Also, increased alpha activity at frontal recording sites has been shown to reflect a state of heightened internal processing demands by inhibiting sensory (bottom-up) information, which may—along with alpha increases in parietal regions (primarily in the right hemisphere)—facilitate the combination of information that is normally widely distributed over the whole brain.

The Role of Individual Differences

Brain correlates underlying creative cognition have been also studied from an individual differences perspective. In this context, individuals high versus low in creativity were compared with respect to functional patterns of brain activity during the performance of different creativity-related tasks (Bhattacharya & Petsche, 2005; Carlsson et al., 2000; Chávez-Eakle et al., 2007; Jaušovec, 2000; Martindale et al., 1984). For instance, the pioneering work by Martindale and Hines (1975) revealed evidence that highly creative individuals were more likely to exhibit higher alpha wave activity while performing the alternate uses test than less creative individuals. Similarly, in Martindale and Hasenfus (1978) highly creative individuals showed higher levels of alpha than less creative subjects during an inspirational phase (e.g., while they had to make up a creative story) but not during an analogue of creative elaboration (e.g., writing down the story). Interestingly, this effect was more pronounced when individuals were explicitly instructed to be original in generating their responses (see Martindale & Hasenfus, 1978). In a more recent study, Jaušovec (2000) also observed evidence that highly creative individuals showed higher EEG alpha power measures than did average creative individuals while they were engaged in the performance of creativity problems.

The findings of our laboratory nicely fit into this picture. In Fink et al. (2009a), we investigated EEG alpha activity while participants were required to generate alternative, original uses of common, everyday objects (e.g., "umbrella," "cap," "pencil," "vase of flowers"). Based on the participants' originality of ideas, the total sample was divided into groups of lower and higher originality. We observed evidence that higher original individuals exhibited a comparatively strong hemispheric asymmetry with respect to alpha activity, with a stronger task-related alpha synchronization in the right than in the left hemisphere, while in less original individuals no hemispheric differences with respect to alpha activity emerged. Similarly, during imagining an improvisational dance, professional dancers exhibited more right-hemispheric alpha synchronization in parietotemporal and parieto-occipital areas than novices (Fink et al., 2009b). This finding is particularly interesting in view of the fact that the creativity-related brain states which we have repeatedly observed in our studies on creative cognition seem not to be restricted to the comparatively basic AU task; rather, they are apparent in more applied creativity-related domains as well.

The Role of Intelligence

A well-established finding in the neuroscientific study of human intelligence is that brighter individuals use their brains more efficiently when engaged in the performance of cognitively demanding tasks than less intelligent people do (Jung & Haier, current volume). This phenomenon, referred to as neural efficiency (Neubauer et al., 2002, 2005; for a recent review see Neubauer & Fink, 2009), has been confirmed in a variety of studies employing a broad range of cognitive task demands. However, some studies in this field of research revealed evidence that neurally efficient brain functioning appears to be moderated by task content and individuals' sex (Neubauer et al., 2002, 2005). Motivated by these findings, which indicate that females and males of varying verbal ability show different patterns of brain activity when engaged in the performance of verbal tasks, Fink and Neubauer (2006) have also investigated sex- and intelligence-related effects on brain activity in the context of creative idea generation (as the production or generation of ideas also falls into the verbal stimulus domain). The findings of this study suggest that males and females of varying verbal intelligence level exhibit different patterns of alpha synchronization, particularly apparent in frontal regions of the cortex. While verbally proficient females (in contrast to those of average verbal intelligence) displayed a stronger synchronization of alpha activity during the production of original ideas, in males the opposite pattern was observed: The production of original ideas in verbally intelligent males was accompanied by a lower synchronization of alpha activity than in the group of average verbal ability. The finding that females and males of varying verbal ability showed different patterns of alpha synchronization during the generation of creative ideas is particularly exciting inasmuch as it resembles the result pattern that we have tentatively also observed on the behavioral level. Analyses of performance data yielded a higher ideational fluency (viewed as a prerequisite of high originality; Guilford, 1950) in verbally intelligent females than in females with average verbal ability, whereas in the male sample exactly the opposite was found, that is, a higher fluency in the group of average than in the group of verbally more intelligent individuals.

Both during the performance of intelligence-related tasks (verbal and visuospatial task employed in Neubauer et al., 2005) and during creative idea generation (Fink & Neubauer, 2006) we observed intelligence and sex-related effects on EEG alpha activity. As has been revealed by the Neubauer et al. (2005) study, during the performance of a verbal matching task only in females were verbal intelligence and ERD negatively correlated (i.e., neural efficiency). In contrast, when males were engaged in the performance of the verbal task, verbally more intelligent males displayed more brain activation than did less intelligent males. With respect to creative idea generation, males and females again displayed a contrary neurophysiological result pattern as it was evident by different patterns of alpha synchronization in women and men. Though these latter findings certainly await replication in larger samples, they could, along with the Neubauer et al. (2005) findings, point to some interesting sex differences in the processing of verbal stimulus material. Presumably as a result of their higher proficiency in this domain, females seem to process verbal information more efficiently than males do. In terms of the alpha inhibition hypothesis (Klimesch et al., 2007), our findings would be compatible with the interpretation that females are more capable of maintaining a state of heightened internal attention or top-down control on internal information processing (i.e., creative idea generation) by inhibiting interference from external input. However, irrespective of the possible meaning of the observed sex differences, the Fink and Neubauer (2006) study provides some preliminary evidence of an interaction between intelligence and creativity on the neurophysiological level (see also Jaušovec, 2000).

Can Creative Thinking Be Trained Effectively?

Given the immanence of creativity in several areas of our everyday life (e.g., in education, pedagogy, science, and industry), research in this field has also addressed the question as to how creativity-related skills can be improved effectively. This has been realized from different perspectives: Krampen (1997), for instance, reports evidence that systematic relaxation exercises were effective in enhancing creative thinking in children and in elderly people. Similarly, So and Orme-Johnson (2001) observed beneficial effects of transcendental meditation techniques on cognition (including creativity) in adolescent school children. From a more cognitive perspective, there are also techniques that aim at improving creativity-related skills by providing specific problem-solving strategies or by activating existing knowledge (see Hany, 2001). In addition to this, neuroscientific studies in this research field also suggest that positive affect or humor might be favorable in the generation of novel, creative ideas. Positive affect is usually induced by giving small, unanticipated rewards to participants or by having them watch funny cartoons or films. Highly relevant literature in this field of research suggests that positive affect has a beneficial influence on cognition and creative problem solving (e.g., cognitive flexibility, verbal fluency, flexibility in thinking, breadth of attentional selection; see Ashby et al., 1999; Rowe et al., 2007). This effect has been explained by referring to increased dopamine levels of the brain (i.e., stimulation of the reward centers of the brain). Recent neuroimaging studies substantiate this view. For instance, using fMRI, Mobbs et al. (2003) report evidence that humor in response to funny cartoons appears to modulate (along with regions of the cerebral cortex) subcortical brain regions that are associated with the dopaminergic reward centers of the brain (such as the ventral tegmental area or the nucleus accumbens). The effects of humor and the comprehension of puns or jokes are also seen in close relation to brain activity in the right hemisphere (Coulson & Williams, 2005), which likewise plays a crucial role in creative cognition (Bowden et al., 2005; Fink et al., 2009a,b; Grabner et al., 2007; Jung-Beeman, 2005).

The findings briefly reported so far provide some evidence that creativity (or in a broader sense cognition) can be improved by positive affect or techniques such as transcendental meditation or relaxation exercises. However, the vast majority of interventions that are reported in relevant literature are creativity trainings designed to specifically enhance the ability to think divergently. Scott et al. (2004a) have reported a meta-analysis including seventy studies on the efficacy of such trainings and observed an overall effect size of Cohen's $\Delta = 0.64$ (see also Hany, 2001; Lipsey & Wilson, 1993; Rose & Lin, 1984). Additional analyses (Scott et al., 2004b) revealed that more cognitive-oriented training procedures proved to be particularly effective, whereas other commonly applied techniques such as imagery training turned out to be less effective.

We have also reported on a cognitive-oriented, computerized creative thinking training (Benedek et al., 2006). In a subsequent study (Fink et al., 2006) the efficacy of the employed creative thinking training was examined on the neurophysiological level by investigating training-induced changes in brain activity from the pre- to the post-test. We briefly describe these two training studies in the following section.

Training of Creative Thinking by Means of Computerized Divergent Thinking Exercises

Computers may not only be supportive in engineering or in the computational domain, they may also be helpful in creative work (Lubart, 2005). This becomes obvious, for instance, when a number of people (e.g., from different institutions or countries) are conjointly engaged in the production or development of new concepts or ideas. Given that traditional face-to-face creativity techniques (such as brainstorming) are usually quite difficult to realize (e.g., due to limited time and location), computer-based techniques such as electronic brainstorming (EBS—virtual brain storming with participants communicating via computers rather than face-to-face; see, e.g., DeRosa et al., 2007) represent a powerful alternative.

The application of computerized techniques in the particular context of improving creativity-related skills may be associated with several important advantages. As outlined by Benedek et al. (2006), conventional trainings are quite time-consuming and extensive as they require a large number of sessions and often necessitate the presence of a trainer or moderator. By contrast, computerized trainings can be conducted efficiently and in a highly economical way (at any time and place). They enable the experimenter to monitor and objectively document the progress in the course of the training (e.g., the time attended to the training or to specific exercises, successfully completed items, number of produced ideas) and thus obtain information on relevant variables that may explain potential training effects. In addition, a computerized creativity training also provides a setting in which idea generation is not prone to process losses (such as production blocking, social loafing, evaluation apprehension) that are typically observed in face-to-face setting (see Diehl & Stroebe, 1987).

The training employed by Benedek et al. (2006) was provided as software stored on a compact disc, along with a manual (including an installation guide) and a training schedule. The training program required participants to generate creative ideas to a broad range of verbal divergent thinking tasks. Some of the employed training exercises were adopted from well-known creativity tests, while other training exercises were constructed by the authors. Guided by the observation that the creativity tasks in the verbal domain may differ considerably with respect to their task demands, two training variants were realized: (1) In the *verbal* creativity training participants were requested to generate original linguistic products, such as finding slogans (e.g., slogans for the new product "orange-ice"), producing nicknames (e.g., for "coffee"), or generating sentences with three given stimulus words (e.g., "car—fish—book"). (2) By contrast, in the more abstract *functional* creativity training, tasks were included that focused on characteristics and functional relations of objects and situations, such as basic features (e.g., think of the basic features of an "apple"), product improvements (e.g., how could a "bicycle" be improved?), or finding explanations and consequences of given situations (e.g., "what would be the consequences of a new ice age?"). Each training version consisted of seventy-two exercises, which were organized into nine training units, each taking approximately half an hour. Individuals who participated in the training were instructed to complete not more than one training unit per day, but not less than one unit within two days, which results in a total training duration of approximately two weeks.

The efficacy of the training was investigated in a pre- and post-test design. As a measure of creativity, we used two parallel versions of the "Verbaler Kreativitäts-Test" by Schoppe (1975). In general, the results of the Benedek et al. (2006) study suggest that creative thinking (as operationalized by means of ideational fluency) can be enhanced by means of divergent thinking exercises. Analyses moreover revealed that the verbal creativity training, which focused on the generation of creative linguistic ideas, was more effective than the more abstract and difficult functional creativity training. Detailed analyses of the training protocols revealed that the verbal creativity training generally resulted in higher ideation rates than the functional creativity training, and only the former training showed a significant increase of the ideation rate over the course of the training period. According to the self-reports of the participants, the functional creativity training was less interesting and entertaining (in the second half of the training) and somewhat more exhausting than the verbal training, which could have resulted in a decrease in training motivation, thereby negatively affecting performance in the post-test.

The findings obtained in the Benedek et al. (2006) study support the view that (at least some types of) creative cognition can be improved effectively. Also, the study points to the usefulness of computerized techniques in the particular context of improving creativity-related skills. In Fink et al. (2006), we addressed the research question as to how brain activity may change as a consequence of a creative thinking training. In this study we employed only the verbal creativity training, which proved to be more effective than the functional creativity training (Benedek et al., 2006). We measured task-related changes in EEG alpha activity in an EEG pre-test and an EEG post-test while participants were engaged in the performance of different creative idea generation tasks (IS, US, AU, and WE task; see above). Between the pre- and post-test, half of the participants received a verbal creative thinking training, whereas the remaining participants received training only after the post-test (waiting control group). The training procedure was the same as in Benedek et al. (2006).

On the behavioral level, the creative thinking training has proven to effectively enhance the originality of ideas in three (i.e., in the IS, US, and

AU tasks) out of the four creative idea generation tasks. In the more convergent or intelligence-related WE task, which has been shown to be significantly correlated with verbal intelligence (Fink et al., 2006, 2007), however, no training effects were observed. The most important finding of the Fink et al. (2006) study was that training effects were also reflected at the level of the brain. As was evident by a significant interaction between training group and cortical area in the post-test, the training group exhibited a stronger task-related synchronization of frontal alpha activity than the control group. The observed frontal alpha synchronization during creative thinking could point to the possibility that during the generation of novel, original ideas (which presumably also requires top-down processing), frontal brain regions must not become disturbed by interfering cognitive processes as long as ongoing idea generation takes place. Thus, the stronger frontal alpha synchronization due to the creativity training (Fink et al., 2006) could indicate that the participants were effectively trained to keep their attention highly focused on relevant aspects of the task (i.e., the production of novel, original ideas) by suppressing interference from task-irrelevant external input (see Klimesch et al., 2007; Sauseng et al., 2005). Though the functional significance of EEG alpha synchronization in the particular context of creative thinking certainly needs to be addressed in more detail in future studies, the obtained findings are consistent with the view that frontal cortical areas are critically involved in the production of novel, original ideas (e.g., Carlsson et al., 2000; Dietrich, 2004; Flaherty, 2005; Folley & Park, 2005; Goel & Vartanian, 2005; Heilman et al., 2003; Howard-Jones et al., 2005). From a more general perspective, the Fink et al. (2006) study also demonstrates the usefulness of neuroscientific measurement methods in the validation of cognitive trainings and generally points—along with the Benedek et al. (2006) study—to the applicability of computer-based training procedures in the particular context of improving creativity-related skills.

Enhancing Creativity by Means of Cognitive Stimulation

Relevant literature in this field also suggests that creative cognition might be improved by means of cognitive stimulation (e.g., Dugosh et al., 2000). This could be simply realized, for instance, by means of divergent thinking exercises preceding the actual idea-generation task (Coskun, 2005). In addition to this, cognitive stimulation can be also achieved by confronting people with the ideas of others. As is the case in classic group-based brainstorming techniques (Osborn, 1957), each single idea or solution a person generates for a specific problem may stimulate new ideas or solutions in others. In this context, Dugosh and Paulus (2005; see also Dugosh et al., 2000) report exciting empirical findings that the number of generated unique ideas may be enhanced through the exposure to ideas of others (provided that the individuals actively attend to the presented ideas; see also Paulus & Yang, 2000). In two fMRI experiments of our laboratory we addressed the research question of how creative cognition can be improved effectively by means of such types of interventions and whether any intervention effects are also reflected at the level of the brain. Similarly to our previous EEG studies, we presented everyday objects (such as a "tin" or an "umbrella") during fMRI recording and participants were instructed to generate creative or original uses of the given objects, which had to be verbalized by the participants subsequent to a so-called idea-generation phase (see Fink et al., 2010, 2011). The oral responses were recorded by the experimenter and rated with respect to their originality subsequent to the fMRI recording session. In one experimental condition, participants performed the AU task subsequent to a short cognitive stimulation condition during which they were exposed to ideas produced by other people (as they were obtained in a pre-experimental pilot study). The findings of the Fink et al. (2010) study reveal performance increases as a result of the employed creativity interventions, and more importantly, effects were also apparent at the level of the brain. The employed interventions recruit a complex and widespread neural network primarily involving posterior brain regions that are known as important components of the neural network specialized for semantic information processing.

Concluding Remarks

Neuroscientific studies on creative cognition have revealed valuable insights into potential brain mechanisms that underlie various facets of creative cognition. For instance, research has shown that brain activity in response to more divergent or creativity-related tasks (such as responding creatively to hypothetical or utopian situations) differ from brain activity patterns during the performance of more convergent or intelligence-related tasks (such as completing given words or performing mental arithmetic; Fink et al., 2006, 2007; Jaušovec & Jaušovec, 2000; Mölle et al., 1999; Razumnikova, 2000). Studies on creative cognition have also yielded evidence that brain states accompanying highly original ideas differ from those observed during the production of less original, conventional ideas (as determined by external or subjective ratings; Fink & Neubauer, 2006;

Grabner et al., 2007; see also research on the subjective experience of "*aha*!," Jung-Beeman et al., 2004). From an individual differences perspective we could—in continuation of our work on neural efficiency (Neubauer et al., 2002, 2005)—also demonstrate that the production of original ideas seems to be moderated by participants' sex and intelligence level (Fink & Neubauer, 2006) and by individual differences in the personality dimension of extraversion–introversion (Fink & Neubauer, 2008). Finally, research in this field also suggests that creative cognition can be improved effectively by means of training (Benedek et al., 2006) or cognitive stimulation (Fink et al., 2010, 2011) and that performance increases can be validated at the neurophysiological level (Fink et al., 2006; 2010, 2011).

It is worth noting that EEG activity in the alpha frequency band has proven to be fairly sensitive to creativity-related demands in a series of studies (for a review, see Fink & Benedek, 2013). Specifically, on the basis of existing evidence on the relationship between EEG alpha activity and creative cognition we can conclude that EEG alpha activity varies as a function of the creative demands of a task (the more creative a task, the higher the level of alpha activity; Fink et al., 2007), as a function of originality (higher originality is accompanied by more alpha; Fink & Neubauer, 2006; Grabner et al., 2007) or subjective experience of insight (more alpha in insight vs. noninsight solutions; Jung-Beeman et al., 2004), and as a function of an individual's creativity level (more alpha in higher creative individuals; Fink et al., 2009a; Jaušovec, 2000; Martindale & Hines, 1975). Alpha synchronization has traditionally been considered as a functional correlate of cortical idling, presumably reflecting a reduced state of active information processing in the underlying neuronal networks (Pfurtscheller et al., 1996). However, in the meantime more and more studies suggest that synchronization of alpha activity does not merely reflect cortical deactivation or cortical idling (a highly readable review on this topic is given in Klimesch et al., 2007). In fact, alpha synchronization appears to be especially relevant during internal processing demands, for instance when participants are required to hold information temporarily in mind (see Sauseng et al., 2005). Along these lines, the diffuse and topographically less clear pattern of alpha synchronization in posterior parietal brain regions, which we have repeatedly observed in our studies on creative cognition (e.g., Fink et al., 2007, 2009a,b), could reflect the absence of stimulus-driven, external bottom-up stimulation and, thus, a form of top-down activity (Benedek et al., 2011; von Stein & Sarnthein, 2000) or a state of heightened internal attention facilitating the (re-)combination of semantic information that is normally distantly related.

Though the findings summarized in this chapter may uncover some brain correlates underlying creative cognition, some important issues remain unresolved. First and foremost, the employed creativity tasks used in neuroscientific studies on creative cognition are essentially basic types of tasks, which had to be modified in order to be reasonably applicable in EEG or fMRI measurements. In this particular context it can be argued that the employed tasks are too simple to be generalizable to "real-life" creative achievements. The difficulty of operationalizing creativity in neuroscientific studies of creative cognition is additionally complicated by the fact that participants are required to be creative while they are mounted with an electrode cap sitting in a shielded EEG cabin or lying supine in the fMRI scanner. Thus, future neuroscientific research on creativity may not only be challenged by the investigation of brain activity in tasks with valid psychometric properties (Arden et al., 2010), but also in more complex, ecologically valid "real-life" creativity tasks. Promising examples for this exciting new research line include the studies of Berkowitz and Ansari (2010), Bhattacharya and Petsche (2005), and Kowatari et al. (2009), who extend neuroscientific research to the domain of artistic creativity including the study of brain activity during musical improvisation, visual art, and designing new pens, respectively (for a recent EEG study on dance improvisation see Fink et al., 2009b). On the other hand, however, it has also been argued that the employed tasks might be too complex, and thus do not allow us to link the evidence with single definable neurocognitive processes (e.g., Dietrich & Kanso, 2010). That is, the neuroscientific research on creativity might also benefit from the employment of simpler tasks and paradigms, which can more easily be related to well-established concepts of cognitive neuroscience such as attention, memory, or cognitive control. This approach would thus not study creativity as a unitary construct, but would study relevant aspects of it, thereby trying to promote neurocognitive theories of creativity.

Perhaps the most important benefit of the summarized research on creative cognition is that it may also entail some relevant practical implications. The work presented in this chapter not only reveals some valuable brain correlates underlying creative cognition, it moreover suggests that at least some facets of creative cognition can be trained or stimulated effectively and that the effects of such interventions are also observable at the level of the brain. This could be viewed as a highly promising objective in the field of cognition inasmuch as relevant research not only focuses on describing the status quo of an individual in a particular variable of interest (such as intelligence or creativity) but also adopts a dynamic view of cognition that incorporates the crucial importance of learning or training in the course of expertise acquisition in a particular cognitive domain. Meanwhile, neuroscientific studies have accumulated a large body of empirical evidence substantiating this view. For instance, research has revealed that training of reasoning (Neubauer et al., 2004, 2010), mental arithmetic (Ischebeck et al., 2006), creative cognition (Fink et al., 2006), and the treatment of orthographic spelling in dyslexic children (Richards et al., 2006; Weiss et al., 2010) are accompanied by specific changes in activity patterns of the brain (for training-induced changes of structural parameters of the brain see, e.g., Maguire et al., 2000; Mechelli et al., 2004; Münte et al., 2002).

The enhancement of creativity-related skills may also be a fruitful avenue for future research. Progress in the scientific understanding of how creative cognition can be enhanced involves important practical implications, particularly for the pedagogical or educational domain. In light of the view that the "plastic" brain is sensitive to environmental stimulation (see Garlick, 2002; see also Münte et al.'s 2002 report on "the musician's brain as a model of neuroplasticity"), we are all—practitioner and scientists—challenged to attend to the question of how the cognitive capacities of an individual can be realized to the best possible extent.

References

Amabile, T. M. (1982). Social psychology of creativity: A consensual assessment technique. *Journal of Personality and Social Psychology, 43*, 997–1013.

Amabile, T. M. (1983). Social psychology of creativity: A componential conceptualization. *Journal of Personality and Social Psychology, 45*, 357–376.

Arden, R., Chavez, R. S., Grazioplene, R., & Jung, R. E. (2010). Neuroimaging creativity: A psychometric review. *Behavioural Brain Research, 214*, 143–156.

Ashby, F. G., Isen, A. M., & Turken, A. U. (1999). A neuropsychological theory of positive affect and its influence on cognition. *Psychological Review, 106*, 529–550.

Bechtereva, N. P., Korotkov, A. D., Pakhomov, S. V., Roudas, M. S., Starchenko, M. G., & Medvedev, S. V. (2004). PET study of brain maintenance of verbal creative activity. *International Journal of Psychophysiology, 53*, 11–20.

Benedek, M., Bergner, S., Könen, T., Fink, A., & Neubauer, A. C. (2011). EEG alpha synchronization is related to top-down processing in convergent and divergent thinking. *Neuropsychologia, 49*, 3505–3511. doi:10.1016/j.neuropsychologia.2011.09.004.

Benedek, M., Fink, A., & Neubauer, A. C. (2006). Enhancement of ideational fluency by means of computer-based training. *Creativity Research Journal, 18,* 317–328.

Berkowitz, A. L., & Ansari, D. (2010). Expertise-related deactivation of the right temporoparietal junction during musical improvisation. *NeuroImage, 49,* 712–719.

Bhattacharya, J., & Petsche, H. (2005). Drawing on mind's canvas: Differences in cortical integration patterns between artists and non-artists. *Human Brain Mapping, 26,* 1–14.

Bowden, E. M., Jung-Beeman, M., Fleck, J., & Kounios, J. (2005). New approaches to demystifying insight. *Trends in Cognitive Sciences, 9,* 322–328.

Carlsson, I., Wendt, P. E., & Risberg, J. (2000). On the neurobiology of creativity: Differences in frontal activity between high and low creative subjects. *Neuropsychologia, 38,* 873–885.

Chávez-Eakle, R. A., Graff-Guerrero, A., García-Reyna, J., Vaugier, V., & Cruz-Fuentes, C. (2007). Cerebral blood flow associated with creative performance: A comparative study. *NeuroImage, 38,* 519–528.

Coskun, H. (2005). Cognitive stimulation with convergent and divergent thinking exercises in brainwriting: Incubation, sequence priming, and group context. *Small Group Research, 36,* 466–498.

Coulson, S., & Williams, S. (2005). Hemispheric asymmetries and joke comprehension. *Neuropsychologia, 43,* 128–141.

DeRosa, D. M., Smith, C. L., & Hantula, D. A. (2007). The medium matters: Mining the long-promised merit of group interaction in creative idea generation tasks in a meta-analysis of the electronic group brainstorming literature. *Computers in Human Behavior, 23,* 1549–1581.

Diehl, M., & Stroebe, W. (1987). Productivity loss in brainstorming groups: Toward the solution of a riddle. *Journal of Personality and Social Psychology, 53,* 497–509.

Dietrich, A., & Kanso, R. (2010). A review of EEG, ERP, and neuroimaging studies of creativity and insight. *Psychological Bulletin, 136,* 822–848.

Dietrich, A. (2004). The cognitive neuroscience of creativity. *Psychonomic Bulletin & Review, 11,* 1011–1026.

Dietrich, A. (2007). Who's afraid of a cognitive neuroscience of creativity? *Methods, 42,* 22–27.

Dugosh, K. L., & Paulus, P. B. (2005). Cognitive and social comparison processes in brain storming. *Journal of Experimental Social Psychology, 41,* 313–320.

Dugosh, K. L., Paulus, P. B., Roland, E. J., & Yang, H.-C. (2000). Cognitive stimulation in brainstorming. *Journal of Personality and Social Psychology, 79,* 722–735.

Feist, G. J. (1998). A meta-analysis of personality in scientific and artistic creativity. *Personality and Social Psychology Review, 2,* 290–309.

Fink, A., & Benedek, M. (2013). EEG Alpha power and creative ideation. *Neuroscience and Biobehavioral Reviews.* Advance online publication. doi:10.1016/j.neubiorev.2012.12.002.

Fink, A., & Neubauer, A. C. (2006). EEG alpha oscillations during the performance of verbal creativity tasks: Differential effects of sex and verbal intelligence. *International Journal of Psychophysiology, 62,* 46–53.

Fink, A., & Neubauer, A. C. (2008). Eysenck meets Martindale: The relationship between extraversion and originality from the neuroscientific perspective. *Personality and Individual Differences, 44,* 299–310.

Fink, A., Benedek, M., Grabner, R. H., Staudt, B., & Neubauer, A. C. (2007). Creativity meets neuroscience: Experimental tasks for the neuroscientific study of creative thinking. *Methods, 42,* 68–76.

Fink, A., Grabner, R. H., Benedek, M., & Neubauer, A. C. (2006). Divergent thinking training is related to frontal electroencephalogram alpha synchronization. *European Journal of Neuroscience, 23,* 2241–2246.

Fink, A., Grabner, R. H., Benedek, M., Reishofer, G., Hauswirth, V., Fally, M., et al. (2009a). The creative brain: Investigation of brain activity during creative problem solving by means of EEG and fMRI. *Human Brain Mapping, 30,* 734–748.

Fink, A., Grabner, R. H., Gebauer, D., Reishofer, G., Koschutnig, K., & Ebner, F. (2010). Enhancing creativity by means of cognitive stimulation: Evidence from an fMRI study. *NeuroImage, 52,* 1687–1695.

Fink, A., Graif, B., & Neubauer, A. C. (2009b). Brain correlates underlying creative thinking: EEG alpha activity in professional vs. novice dancers. *NeuroImage, 46,* 854–862.

Fink, A., Koschutnig, K., Benedek, M., Reishofer, G., Ischebeck, A., Weiss, E. M., et al. (2011). Stimulating creativity via the exposure to other people's ideas. *Human Brain Mapping.* doi:10.1002/hbm.21387.

Flaherty, A. W. (2005). Frontotemporal and dopaminergic control of idea generation and creative drive. *Journal of Comparative Neurology, 493,* 147–153.

Folley, B. S., & Park, S. (2005). Verbal creativity and schizotypal personality in relation to prefrontal hemispheric laterality: A behavioral and near-infrared optical imaging study. *Schizophrenia Research, 80,* 271–282.

Garlick, D. (2002). Understanding the nature of the general factor of intelligence: The role of individual differences in neural plasticity as an explanatory mechanism. *Psychological Review, 109,* 116–136.

Goel, V., & Vartanian, O. (2005). Dissociating the roles of right ventral lateral and dorsal lateral prefrontal cortex in generation and maintenance of hypotheses in set-shift problems. *Cerebral Cortex, 15,* 1170–1177.

Grabner, R. H., Fink, A., & Neubauer, A. C. (2007). Brain correlates of self-rated originality of ideas: Evidence from event-related power and phase-locking changes in the EEG. *Behavioral Neuroscience, 121,* 224–230.

Guilford, J. P. (1950). Creativity. *American Psychologist, 5,* 444–454.

Guilford, J. P. (1967). *The nature of human intelligence*. New York: McGraw-Hill.

Hany, E. A. (2001). Förderung von Kreativität. In K. J. Klauer (Ed.), *Handbuch Kognitives Training* (pp. 261–291). Göttingen: Hogrefe.

Heilman, K. M., Nadeau, S. E., & Beversdorf, D. O. (2003). Creative innovation: possible brain mechanisms. *Neurocase, 9,* 369–379.

Hennessey, B. A., & Amabile, T. M. (2010). Creativity. *Annual Review of Psychology, 61,* 569–598.

Howard-Jones, P. A., Blakemore, S.-J., Samuel, E. A., Summers, I. R., & Claxton, G. (2005). Semantic divergence and creative story generation: An fMRI investigation. *Cognitive Brain Research, 25,* 240–250.

Ischebeck, A., Zamarian, L., Siedentopf, C., Koppelstätter, F., Benke, T., Felber, S., et al. (2006). How specifically do we learn? Imaging the learning of multiplication and subtraction. *NeuroImage, 30,* 1365–1375.

Jauk, E., Benedek, M., & Neubauer, A. C. (2012). Tackling creativity at its roots: Evidence for different patterns of EEG alpha activity related to convergent and divergent modes of task processing. *International Journal of Psychophysiology, 84,* 219–225. doi:10.1016/j.ijpsycho.2012.02.012.

Jaušovec, N., & Jaušovec, K. (2000). EEG activity during the performance of complex mental problems. *International Journal of Psychophysiology, 36,* 73–88.

Jaušovec, N. (2000). Differences in cognitive processes between gifted, intelligent, creative, and average individuals while solving complex problems: An EEG Study. *Intelligence, 28,* 213–237.

Jung, R. E., Segall, J. M., Jeremy Bockholt, H., Flores, R. A., Smith, S. M., Chavez, R. S., et al. (2010). Neuroanatomy of creativity. *Human Brain Mapping, 31,* 398–409.

Jung-Beeman, M. (2005). Bilateral brain processes for comprehending natural language. *Trends in Cognitive Sciences, 9,* 512–518.

Jung-Beeman, M., Bowden, E. M., Haberman, J., Frymiare, J. L., Arambel-Liu, S., Greenblatt, R., et al. (2004). Neural activity when people solve verbal problems with insight. *PLoS Biology, 2,* 500–510.

Kaufman, J. C. (2005). The door that leads into madness: Eastern European poets and mental illness. *Creativity Research Journal, 17*, 99–103.

King, L. A., Walker, L. M., & Broyles, S. J. (1996). Creativity and the Five-Factor model. *Journal of Research in Personality, 30*, 189–203.

Klimesch, W. (1999). EEG alpha and theta oscillations reflect cognitive and memory performance: A review and analysis. *Brain Research Reviews, 29*, 169–195.

Klimesch, W., Sauseng, P., & Hanslmayr, S. (2007). EEG alpha oscillations: The inhibition-timing hypothesis. *Brain Research Reviews, 53*, 63–88.

Kowatari, Y., Lee, S. H., Yamamura, H., Nagamori, Y., Levy, P., Yamane, S., et al. (2009). Neural networks involved in artistic creativity. *Human Brain Mapping, 30*, 1678–1690.

Krampen, G. (1997). Promotion of creativity (divergent productions) and convergent productions by systematic-relaxation exercises: Empirical evidence from five experimental studies with children, young adults, and elderly. *European Journal of Personality, 11*, 83–99.

Lachaux, J.-P., Rodriguez, E., Martinerie, J., & Varela, F. J. (1999). Measuring phase synchrony in brain signals. *Human Brain Mapping, 8*, 194–208.

Lipsey, M. W., & Wilson, D. B. (1993). The efficacy of psychological, educational, and behavioral treatment: Confirmation from meta-analysis. *American Psychologist, 48*, 1181–1209.

Lubart, T. (2005). How can computers be partners in the creative process: Classification and commentary on the special issue. *International Journal of Human–Computer Studies, 63*, 365–369.

Maguire, E. A., Gadian, D. G., Johnsrude, I. S., Good, C. D., Ashburner, J., Frackowiak, R. S., et al. (2000). Navigation-related structural change in the hippocampi of taxi drivers. *Proceedings of the National Academy of Sciences of the United States of America, 97*, 4398–4403.

Martindale, C., Hines, D., Mitchell, L., & Covello, E. (1984). EEG alpha asymmetry and creativity. *Personality and Individual Differences, 5*, 77–86.

Martindale, C., & Hasenfus, N. (1978). EEG differences as a function of creativity, stage of the creative process, and effort to be original. *Biological Psychology, 6*, 157–167.

Martindale, C., & Hines, D. (1975). Creativity and cortical activation during creative, intellectual, and EEG feedback tasks. *Biological Psychology, 3*, 71–80.

Mechelli, A., Crinion, J. T., Noppeney, U., O'Dohorty, J., Ashburner, J., Frackowiak, R. S., et al. (2004). Structural plasticity in the bilingual brain. *Nature, 431*, 757.

Mobbs, D., Greicius, M. D., Abdel-Azim, E., Menon, V., & Reiss, A. L. (2003). Humor modulates the mesolimbic reward centres. *Neuron, 40*, 1041–1048.

Mölle, M., Marshall, L., Wolf, B., Fehm, H. L., & Born, J. (1999). EEG complexity and performance measures of creative thinking. *Psychophysiology, 36*, 95–104.

Münte, T. F., Altenmüller, E., & Jäncke, L. (2002). The musician's brain as a model of neuroplasticity. *Nature Reviews: Neuroscience, 3*, 473–478.

Neubauer, A. C., Bergner, S., & Schatz, M. (2010). Two- vs. three-dimensional presentation of mental rotation tasks: Sex differences and effects of training on performance and brain activation. *Intelligence, 38*, 529–539.

Neubauer, A. C., & Fink, A. (2009). Intelligence and neural efficiency. *Neuroscience and Biobehavioral Reviews, 33*, 1004–1023.

Neubauer, A. C., Fink, A., & Schrausser, D. G. (2002). Intelligence and neural efficiency: The influence of task content and sex on the brain-IQ relationship. *Intelligence, 30*, 515–536.

Neubauer, A. C., Grabner, R. H., Fink, A., & Neuper, C. (2005). Intelligence and neural efficiency: Further evidence of the influence of task content and sex on the brain-IQ relationship. *Cognitive Brain Research, 25*, 217–225.

Neubauer, A. C., Grabner, R. H., Freudenthaler, H. H., Beckmann, J. F., & Guthke, J. (2004). Intelligence and individual differences in becoming neurally efficient. *Acta Psychologica, 116*, 55–74.

Osborn, A. F. (1957). *Applied imagination*. New York: Scribner's.

Paulus, P. B., & Yang, H.-C. (2000). Idea generation in groups: A basis for creativity in organizations. *Organizational Behavior and Human Decision Processes, 82*, 76–87.

Petsche, H. (1996). Approaches to verbal, visual, and musical creativity by EEG coherence analysis. *International Journal of Psychophysiology, 24*, 145–159.

Pfurtscheller, G. (1999). Quantification of ERD and ERS in the time domain. In G. Pfurtscheller & F. H. Lopes da Silva (Eds.), *Event-related desynchronization: Handbook of electroencephalography and clinical neurophysiology* (Rev. Ed., Vol. 6, pp. 89–105). Amsterdam: Elsevier.

Pfurtscheller, G., Stancak, A., Jr., & Neuper, C. (1996). Event-related synchronization (ERS) in the alpha band—an electrophysiological correlate of cortical idling: a review. *International Journal of Psychophysiology, 24*, 39–46.

Razumnikova, O. M. (2000). Functional organization of different brain areas during convergent and divergent thinking: An EEG investigation. *Cognitive Brain Research, 10*, 11–18.

Razumnikova, O. M. (2004). Gender differences in hemispheric organization during divergent thinking: An EEG investigation in human subjects. *Neuroscience Letters, 362*, 193–195.

Richards, T. L., Aylward, E. H., Berninger, V. W., Field, K. M., Grimme, A. C., Richards, A. L., et al. (2006). Individual fMRI activation in orthographic mapping and morpheme mapping after orthographic or morphological spelling treatment in child dyslexics. *Journal of Neurolinguistics, 19*, 56–86.

Rose, L. H., & Lin, H. J. (1984). A meta-analysis of long-term creativity training programs. *Journal of Creative Behavior, 18*, 11–22.

Rowe, G., Hirsh, J. B., & Anderson, A. K. (2007). Positive affect increases the breadth of attentional selection. *Proceedings of the National Academy of Sciences of the United States of America, 104*, 383–388.

Sandkühler, S., & Bhattacharya, J. (2008). Deconstructing insight: EEG correlates of insightful problem solving. *PLoS ONE, 3*(1), e1459.

Sauseng, P., Klimesch, W., Doppelmayr, M., Pecherstorfer, T., Freunberger, R., & Hanslmayr, S. (2005). EEG alpha synchronization and functional coupling during top-down processing in a working memory task. *Human Brain Mapping, 26*, 148–155.

Sawyer, R. K. (2006). Educating for innovation. *Thinking Skills and Creativity, 1*, 41–48.

Schoppe, K. (1975). *Verbaler Kreativitäts-Test (V-K-T)*. Göttingen: Hogrefe.

Scott, G., Leritz, L. E., & Mumford, M. D. (2004a). The effectiveness of creativity training: A quantitative review. *Creativity Research Journal, 16*, 361–388.

Scott, G., Leritz, L. E., & Mumford, M. D. (2004b). Types of creativity training: Approaches and their effectiveness. *Journal of Creative Behavior, 38*, 149–179.

Smith, S. M., Ward, T. B., & Finke, R. A. (1995). *The creative cognition approach*. Cambridge, MA: MIT Press.

So, K. T., & Orme-Johnson, D. W. (2001). Three randomized experiments on the longitudinal effects of the transcendental meditation technique on cognition. *Intelligence, 29*, 419–440.

Stein, M. I. (1953). Creativity and culture. *Journal of Psychology, 36*, 311–322.

Sternberg, R. J., & Lubart, T. I. (1996). Investing in creativity. *American Psychologist, 7*, 677–688.

Torrance, E. P. (1966). *Torrance tests of creative thinking*. Bensenville, IL: Scholastic Testing Service.

von Stein, A., & Sarnthein, J. (2000). Different frequencies for different scales of cortical integration: From local gamma to long range alpha/theta synchronization. *International Journal of Psychophysiology, 38*, 301–313.

Ward, T. B. (2007). Creative cognition as a window on creativity. *Methods, 42*, 28–37.

Weiss, S., Grabner, R. H., Kargl, R., Purgstaller, C. & Fink, A. (2010). Behavioral and neurophysiological effects of a computer-aided morphological awareness training on spelling and reading skills. *Reading and Writing: An Interdisciplinary Journal, 23*, 645–671.

11 Creativity and Intelligence: Brain Networks That Link and Differentiate the Expression of Genius

Rex E. Jung and Richard J. Haier

If you're so smart, why aren't you a genius? The simple answer is that intelligence and creativity are not the same thing and genius apparently requires both (Jensen, 1998). Psychology has a long history of discussing this issue, and numerous distinctions have been hypothesized to augment general definitions of intelligence, creativity, and genius. Empirical testing of competing ideas, however, depends on measurement. Tests of intelligence and tests of creativity evolved during the twentieth century and both kinds of tests now have good psychometric qualities. Early research using electroencephalographic (EEG) techniques and positron emission tomography (PET) suggested that (1) intelligence and creativity test scores were related to neural activity (i.e., excitation) in the frontal lobes and other areas (with hemispheric differences), and that (2) less activity (i.e., neuronal disinhibition) was related to higher scores for both concepts (see Runco, 2004, for a review of creativity studies, and see Jung & Haier, 2007, for a review of intelligence studies). Prior to the advent of neuroimaging techniques, Eysenck (1995) formulated a theory of creativity and the brain that proposed the importance of disinhibition (i.e., less neuronal activation), especially in the frontal lobes. Thus, while the interplay of both neuronal excitatory and disinhibitory processes characterize studies of creativity, the focus on neuronal disinhibition, particularly within the frontal lobes, predates the advent of modern neuroimaging studies.

The widespread availability of sophisticated neuroimaging technologies and analysis techniques allows new kinds of studies using psychometric tools to investigate brain characteristics of both concepts. Now we can address more complex questions. The ultimate aim is to determine the specific neural networks that underlie intelligence and creativity, in their various forms and especially as they relate to genius. Two key questions are (1) whether, and to what extent, such networks overlap and (2) what unique aspects of the network must be present which make the simultaneous expression of intelligence and creativity (i.e., genius) relatively rare? These questions and the implications of tentative answers are the focus of this chapter. We will present a brain model for intelligence and a model for creativity and discuss how genius may emerge from the overlapping and unique aspects of these models.

Before we get to the brain models and the studies on which they are based, we need to note two points. First is the emerging distinction, within the cognitive neurosciences, between the association of regions of interest with general cognitive functioning and the identification of brain networks subserving specific cognitive tasks. This is a rather subtle distinction, but it is an important one for appreciating the potential interplay of intelligence and creativity. Thousands of studies implicate various brain regions as "central" to numerous cognitive tasks including working memory, visual attention, episodic memory, and problem solving, to name a few. A review paper of 275 brain imaging paradigms identified similar brain regions activated during performance of such cognitive tasks (Cabeza & Nyberg, 2000). Several interesting generalizations emerged from this article: (1) vastly different cognitive tasks (e.g., space perception, working memory) engaged similar brain regions; (2) the anterior cingulate is engaged during a wide range of "demanding" cognitive tasks involving "intention to act" (or inhibition of action); and (3) contrary to popular belief, specific "brain regions are not committed to specific functions." We point to these statements to illustrate the empirical basis for the shift away from the phrenology-like idea of one brain area for one cognitive function to a perspective that focuses on the many brain areas that work together in a network. While we will be discussing particular regions of interest identified within individual studies, our overarching goal will be to forge a network of prospective regions subserving intelligence, creativity, and genius.

Second, predating neuroimaging, there is a long history within the neurosciences of evaluating brain function through careful examination of case studies and/or lesion analysis. This is due to the fact that, while multiple brain areas might serve a given cognitive function, removal of discrete brain region through disease or injury will reveal brain regions critical to performance of such functions. Three iconic examples include (1) Phineas Gage, who survived the passage of an iron rod through his frontal lobes resulting in profoundly changed personality (Harlow, 1848); (2) "Tan," the description of whom led to the localization of expressive speech areas of the brain (Broca, 1861); and (3) "H.M.," whose bilateral temporal surgical lesions led to heightened understanding of memory encoding (Scoville & Milner, 1957). Less frequently studied, although equally important, were the brains of examples of extreme cognitive ability, such as savant abilities of extraordinary memory or mathematical calculation and synesthesia (blending of senses such as seeing numbers as colors). These studies suggest (1) that savant ability comes at the expense of "executive or integrative (brain) mechanisms" (Snyder & Mitchell, 1999) and (2) that artistic and musical ability may appear suddenly after left temporal degeneration (Miller, Boone, Cummings, Read, & Mishkin, 2000) and synesthesia may appear after brain damage (Ro, et al., 2007).

In this chapter, we will review these case studies in more detail and then review the brain imaging literature relevant to intelligence and creativity. Our goal is to develop comparative brain models. With a few exceptions, these reports address either creativity or intelligence. However, we start with the one major example in the neurological annals who provides important clues regarding the trifecta of intelligence, creativity, and genius: Albert Einstein.

Intelligence

A Case Report of Genius

Albert Einstein is perhaps the most revered intellectual icon of the twentieth century, and one of the few figures of human progress for whom we have retained the brain for scientific study. In his "golden year" he produced four stunning papers, covering Brownian motion (Einstein, 1905b), the special theory of relativity (Einstein, 1905d), statistical mechanics (Einstein, 1905a), and the photoelectric effect (Einstein, 1905c), out of which his most famous postulate emerged ($E = mc^2$), arguably the most recognizable formula representing applied genius in human history. He was sympathetic to the notion of scientific research on his brain, and underwent EEG recordings during his life. Einstein died on April 18, 1955, at the age of 76 from a ruptured aortic aneurysm, mentally adept to the end. His brain was removed on the morning of his death by Thomas S. Harvey, a pathologist at Princeton Hospital, with the consent of the family (Brian, 1996) and the estate executor (Highfield & Carter, 1993). The brain was photographed, measured with calipers, weighed, fixed in formalin for several months, and subsequently sectioned into about 240 blocks, each consisting of 10 cm^3 of tissue, and embedded in celloidin (Witelson, Kigar, & Harvey, 1999). The brain was described as unremarkable in appearance, and the weight (1,230 grams), length (17.2 cm left/16.4 cm right), and width (7.5 cm left/7.5 cm right) of the cerebral hemispheres were all within the average range for men his age (Anderson & Harvey, 1996). The travels and travails of Einstein's brain following removal are described elsewhere (Abraham, 2002; Paterniti, 2000), and following several inquiries to Dr. Harvey regarding the results of the analysis of Einstein's brain ("Brain of Einstein continues peregrinations," 1981; "Brain that rocked physics rests in cider box," 1978), other research reports eventually followed.

More detailed morphological characteristics of Einstein's brain were systematically compared to the brains of thirty-five male controls (mean age = 57+/–11; mean full scale intelligence quotient (FSIQ) = 116) possessing normal neurological and psychiatric status (Witelson et al., 1999). While Einstein's brain weight was significantly lower than that of a younger control cohort (1,230 cm versus 1,400 cm), no differences were observed on measures of corpus callosum area, frontal lobe, and temporal lobe morphology. However, the parietal operculum was not present in Einstein, resulting in a larger expanse of the inferior parietal lobule, extending some 15 percent wider than similar regions of the controls. This unique morphology, found in none of the control subjects, resulted in a supramarginal gyrus undivided by a major sulcus (figure 11.1). The inferior parietal lobule is associated with visuospatial cognition, mathematical reasoning, and imagery of movement (Crammond, 1997), and its expansion was also noted in other cases of prominent physicists and mathematicians (Spitzka, 1907). The authors of this comparative study of Einstein's brain note that "variation in specific cognitive functions may be associated with the structure of the brain regions mediating those functions" and conclude that the parietal lobule may be implicated in visuospatial intelligence (Witelson et al., 1999, 2152).

Two other studies of Einstein's brain investigated whether differences at the cellular level (e.g., neuron/glia) could explain his genius (Anderson & Harvey, 1996; Diamond, Scheibel, Murphy, & Harvey, 1985). The first study used the blocks of tissue obtained from frontal and parietal regions, bilaterally, comprising superior prefrontal and inferior parietal association cortices. These regions of Einstein's brain were compared to eleven controls ranging in age from 47 to 80 years, obtained from a Veteran's Administration (VA) hospital and fixed in a manner similar to those obtained from Einstein. Cell counts were made of neurons, astrocytes, and oligodendrocytes, from which a neuronal-to-glial ratio was computed. Glial cells provide metabolic support (i.e., nutrition) to neurons. Results indicated that, compared to controls, Einstein had significantly fewer neurons per glial cells in the left inferior parietal cortex, which the authors interpreted to suggest "a response by glial cells to greater neuronal metabolic need"

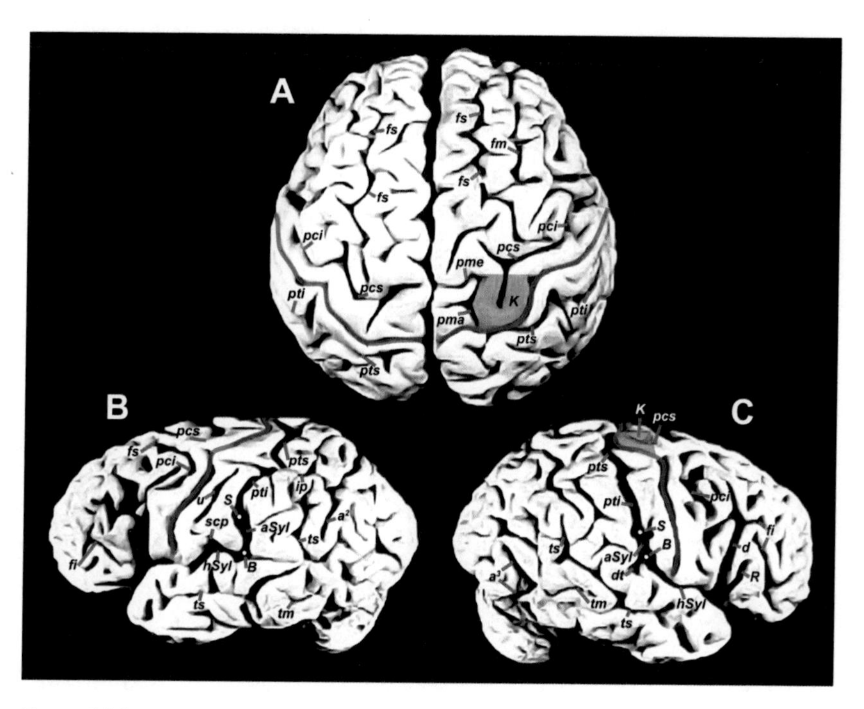

Figure 11.1
Photographs of Einstein's brain (adapted from Falk, 2009). (A) Dorsal view, (B) left lateral view, (C) right lateral view. Sulci: angular (a2), anterior occipital (a3), ascending limb of the posterior Sylvian fissure (aSyl), central fissure (red lines), diagonal (d), descending terminal portion of aSyl (dt), inferior frontal (fi), middle frontal (fm), superior frontal (fs), horizontal limb of the posterior Sylvian fissure (hSyl), intraparietal (ip), precentral inferior and superior (pci, pcs), marginal precentral (pma), medial precentral (pme), postcentral inferior and superior (pti, pts), ascending ramus of Sylvian fissure (R), subcentral posterior sulcus (scp), middle temporal (tm), superior temporal sulcus (ts), unnamed sulcus in postcentral gyrus (u). Other features: branching point between hSyl and aSyl (white dots, B), hand motor cortex knob (K, shaded in A, C), termination of aSyl (white dots, S). With kind permission of Falk, D. (2009). New information about Albert Einstein's brain. *Frontiers in Evolutionary Neuroscience, 1*(3): doi: 10.3389/neuro.18.003.2009.

(Diamond et al., 1985). In other words, Einstein had more glial cells per neuron in this area, suggesting that these neurons might work harder or more efficiently. The second study focused on determining the density of a block of tissue from the right prefrontal association cortex, compared to comparable regions from male controls aged 63 to 79 years. Results indicated increased neuronal packing (i.e., the same number of neurons in a smaller space), which the authors interpret as potentially "decreasing interneuronal conduction time" and thus potentially facilitating cortical connectivity (Anderson & Harvey, 1996). This would be consistent with greater efficiency of brain processing in this region. Thus, these studies suggest that Einstein's brain differed from others in a frontal-parietal network. Information processing in the left inferior parietal lobe (especially the supramarginal gyrus) may have been more powerful and its integration in the right prefrontal cortex may have been more efficient.

Numerous researchers have critiqued the studies of Einstein's brain for various methodological flaws (Galaburda, 1999; Hines, 1998) and, to be sure, the comparison of one exceptional individual to various controls does not lead definitively to localization of genius within the brain. Moreover, not all areas of Einstein's brain have been studied, and a network approach calls for understanding how a key area in the parietal lobe is connected to other areas. We also don't know if the specific brain findings are related to Einstein's intelligence, his creativity, or to his genius.

Brain Networks of Intelligence from Imaging Studies: The P-FIT Theory

People differ in intellectual ability, and these differences are related to features of the brain as determined by neuroimaging studies of the last twenty-five years. We reviewed these studies in 2007; at that time there were thirty-seven studies that used different imaging techniques and different measures of intelligence in samples of different sizes and compositions (Jung & Haier, 2007). Some brain areas were implicated more often than others across these studies. These areas were distributed across the brain but were found mostly in parietal and frontal areas. We proposed the parieto-frontal integration theory (P-FIT) of intelligence, which hypothesized that efficient information flow among these areas, or subgroups of these areas, was a basis for individual differences in intelligence. The P-FIT model is shown in figure 11.2. This hypothesis recognizes that humans gather and process information predominantly through auditory and/or visual means (usually in combination)—thus the network involves a sequence of seven broad information-processing events (shown below in this paragraph in italics). At the start is processing of sensory information

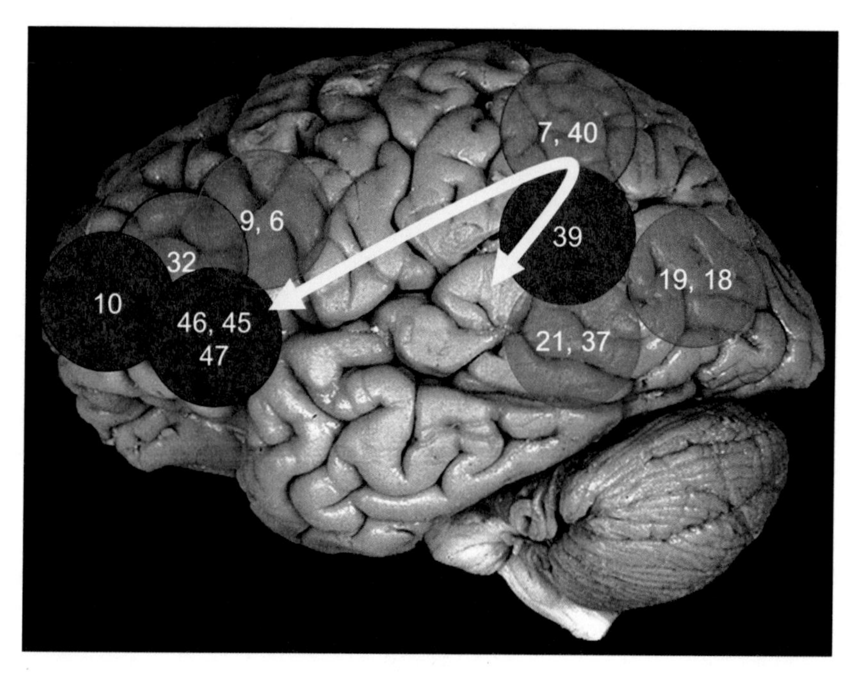

Figure 11.2
P-FIT theory of intelligence. Numbers indicate Brodmann areas. Blue = left lateralized; purple = bilateral; yellow arrow = arcuate fasciculus.

via the extrastriate cortex and fusiform gyrus, involving recognition and subsequent *imagery* of visual input and analysis of auditory *syntax* input in Wernicke's area and surrounding regions. This basic sensory processing is then fed forward to the angular, supramarginal, and inferior parietal cortices, wherein structural *symbolism* and/or *abstraction* are generated and manipulated. The parietal cortex then interacts with frontal regions that serve to *hypothesis test* various solutions to a given problem. The anterior cingulate is involved in *response selection* as well as *inhibition* of competing responses. This process is critically dependent on the fidelity of underlying white matter needed to facilitate rapid and error-free transmission of data between frontal and parietal lobes. The P-FIT is a "network" perspective, conforming to Cabeza and Nyberg's (2000) nascent conceptualization that several brain regions within more integrative "association zones" (i.e., not strictly dedicated to sensory or motor functions) of the frontal and parietal lobes function as cognitive "hubs" subserving multiple cognitive tasks.

Since our 2007 review, more than fifty additional imaging studies of intelligence have been published, most supporting the P-FIT model (Deary,

Penke, & Johnson, 2010). Importantly, these newer studies have become more sophisticated. They have better image-analysis methods (Li, et al., 2009; van den Heuvel, Stam, Kahn, & Pol, 2009), use multivariate combinations of test batteries to extract intelligence factors (Colom et al., 2009), and now commonly have large samples of 100 or more subjects (Tamnes et al., 2010). Developmental studies address P-FIT areas in children and adolescents (Karama et al., 2009; Luders et al., 2011; Schmithorst, 2009), and also are increasingly investigating sex differences (Luders et al., 2008; Tang et al., 2010). Genetic studies have determined that intelligence and brain structures (i.e., gray and white matter) share common genes (Bishop, Fossella, Croucher, & Duncan, 2008; Chiang et al., 2009; Liu et al., 2010). The pace of research publications regarding intelligence and the brain is increasing dramatically (Haier, 2009). These imaging studies of intelligence and the P-FIT, along with the case study reports, offer a framework for studies of creativity that may generate a comparable brain network model.

Creativity

Case Reports Regarding Creativity: Frontotemporal Dementia and Other Lesions

Neurological inquiries regarding creativity converge on the frontal lobes and their inhibitory interactions with temporal, occipital, and parietal lobes (TOP) (Flaherty, 2011; Heilman, Nadeau, & Beversdorf, 2003). This convergence has arisen, at least in part, from several case reports of patients having developed frontotemporal dementia (FTD) and *subsequently* experiencing dramatically increased creative capacity (Miller et al., 1998). Initially, Miller and colleagues reported a few single case reports of creativity in FTD. They subsequently reported that some 17 percent of their entire cohort of sixty-nine patients diagnosed with FTD (twelve patients) exhibited increased visual or musical creativity, and that damage to the left temporal lobe and sparing of the frontal lobes was "a unifying feature of the patients with ability" (Miller et al., 2000, 461). However, left temporal lobe lesions are not exclusively associated with *de novo* artistic expression, which has also been reported in right temporal lobe epilepsy (Mendez, 2005), a case of Parkinson's disease treated with dopaminergic agonists (Schrag & Trimble, 2001), a case of subarachnoid hemorrhage (Lythgoe, Pollak, Kalmus, de Haan, & Chong, 2005), and in a case of insular ischemia (Thomas-Anterion et al., 2010). Subsequent systematic study of artistic ability associated with the various dementias found no general increase in creativity to be linked with FTD (or semantic or dementia of the Alzheimer's type), with the authors noting that "despite the existence of these isolated patients with increased artistic production, however, apathy leading to diminished creativity is more clinically typical of patients with FTD, suggesting that these case studies may be the exception rather than the rule" (Rankin et al., 2007, 49). Thus, these cases suggest that damage to the temporal lobe may be associated with increased artistic creativity in only a very small number of people, but this may be an important clue. Frontal lobe inhibition of the temporal lobe may play a key role, or it may be an alternative pathway to increased creativity.

Neurologist Alice Flaherty has taken a further step in the conceptualization of creative expression with a model of creative drive involving frontotemporal and dopaminergic control of idea generation (Flaherty, 2005). She takes a neurological perspective, focused on the behaviors central to creativity, and weaves together a compelling tapestry from careful study of individual patients. The key features of this model include (1) incorporation of the limbic system as the "driver" of creative pursuits, (2) the notion that creativity is domain independent (i.e., a common component spans creative expressions as varied as artists, scientists, musicians, and so on), and (3) a prediction of similar neurological underpinnings across normal controls, psychiatric patients, and lesion patients. For example, patients with temporal lobe epilepsy were often noted to have a strong drive to write (called "hypergraphia"), also noted in some manic patients, as well as in frontotemporal lobe dementia (FTLD). What these patients had in common was dysfunction of the temporal lobe, which normally inhibits frontal lobe functioning (Menzel et al., 1998). Thus, overt lesions or mild dysfunction to the temporal lobes served to "disinhibit" frontal interactions with other nodes (i.e., language/visuospatial) underlying behavioral output, with right-hemisphere lesions producing higher incidence of hypergraphia, and left lesions producing increased visual and musical output. She also hypothesizes a role for dopamine in novelty-seeking and goal-directed behavior (Mink, 1996). Finally, the frontal lobes are hypothesized to block creative drive when lesioned or dysfunctional (e.g., in depression, anxiety, Wernicke's aphasia). What Flaherty's model introduces to the picture is the notion of mutually inhibitory nodes (i.e., frontal, temporal, subcortical) within a network of brain regions subserving creativity. This model can be tested with neuroimaging.

Neuroimaging of Creativity and the "Frontal Disinhibition Model" (F-DIM)

Brain studies of creativity have not advanced as rapidly as those of intelligence, but results so far are informative and summarized in three recent reviews (Arden, Chavez, Grazioplene, & Jung, 2010; Dietrich & Kanso, 2010; Sawyer, 2011). Arden et al.'s review of forty-five brain-imaging studies of creative cognition did not reveal much consistency among studies. Given the wide range of measures used to assess creativity and the measurement error inherent across the various neuroimaging measures, they conclude that "it is impossible to know whether any results should be attributed to the measures, to the imaging modality or to unreliability in one or both" (152). Dietrich and Kanso reviewed neuroimaging experiments of divergent thinking, artistic creativity, and insight from sixty-three research articles, including the forty-five papers reviewed by Arden et al. (Dietrich & Kanso, 2010). They, too, found that "creative thinking does not appear to depend on any single mental process or brain region, and it is not especially associated with the right hemisphere, defocused attention, low arousal, or alpha synchronization, as sometimes hypothesized" (845). However, they did offer some general conclusions, albeit of a highly qualified nature: "Tasks purportedly involving creative cognition induce changes in prefrontal activity" (ibid.). These changes include both increases and decreases, span all (or most?) of frontal lobe regions, and are not exclusive to the frontal lobes; thus, creativity may not be either "localized" or even "localizable." Sawyer's review of the cognitive neuroscience of creativity similarly notes that (1) "the entire brain is active when people are engaged in creative tasks," (2) "left and right hemispheres are equally activated in most creative tasks," (3) and "the same brain areas are active that are active in many everyday tasks" (149). All three reviews suggest that the construct of "creativity" would benefit greatly from further parsing into subcomponents from which more fine-grained cognitive neurosciences results might emerge. However, all three reviews rely almost exclusively on functional (i.e., EEG, functional magnetic resonance imaging [fMRI]) studies. All functional imaging studies are influenced by task demands during image acquisition. Thus, the inability to localize a network of underlying creativity may have as much to do with methodological vagaries related to task and acquisition techniques as with construct problems. Structural and lesion studies avoid task demand problems.

Luckily for the construct of creativity, divergent thinking has long been parsed into subcomponents comprised of fluency (i.e., the raw number of items produced), flexibility (i.e., different conceptual categories produced), and originality (i.e., novel responses produced). The notion of "originality" permeates the creativity literature (Runco & Charles, 1993), and one recent study provides important insights (Shamay-Tsoory, Adler, Aharon-Peretz, Perry, & Mayseless, 2011). Forty patients with localized brain damage (i.e., lesions) to various regions, and seventeen matched controls, completed the Torrance Test of Creative Thinking and the Alternate Uses Test, both reliable and valid measures of one aspect of divergent thinking. In those subjects with medial frontal lesions, and particularly right medial frontal lesions, originality scores (the "novel" part of "novel and useful") across measures were significantly reduced. Similarly, in those subjects with left parietal lobe lesions, originality scores were significantly higher, even significantly higher than normal control subjects. The authors interpret their findings to support a right lateralized frontoparietal network of brain regions supporting originality, with "lesions in the right hemisphere (being) associated with impaired creativity, whereas damage to the left hemisphere (being) associated with somewhat increased creativity." Taken together with the studies showing left temporal lobe degeneration associated with increased artistic and musical creativity in patients with FTD, this study suggests that *lower* brain integrity within left hemisphere brain structures—particularly left anterior temporal and inferior parietal lobes—serves to "disinhibit" other brain regions associated with increased novelty generation as measured by both artistic endeavors, and psychometric tests of divergent thinking.

We recently completed three "structural" imaging studies of creative cognition that have several advantages compared to the lesion studies reviewed above: (1) they are applied to large (i.e., >50) samples of healthy, young individuals; (2) they use reliable and valid measures of intelligence (Wechsler Scales), creativity (i.e., Alternate Uses Test; Creative Achievement Test), and personality (i.e., NEO-FFI) (Costa & McCrae, 1992); (3) they use neuroimaging measures that are not dependent upon task-related functional changes; and (4) they also assess "originality" as distinct from "fluency" or "flexibility" factors of divergent thinking. In our first study, we probed the relationship between creative cognition and concentration of N-acetyl-aspartate, a marker of neuronal integrity, in a sample of fifty-six healthy people using proton magnetic resonance spectroscopy (MRS) (Jung, Gasparovic, Chavez, Flores et al., 2009). Three divergent thinking tasks (i.e., Alternate Uses Test) were ranked by three judges to create a creativity index using the consensual assessment technique (Amabile, 1982). N-acetyl-aspartate concentration was inversely correlated with creative cognition in the right anterior cingulate for high IQ subjects (>116 FSIQ), but positively correlated with creative cognition in the left anterior cingulate for average IQ subjects (<116 FSIQ). This finding is consistent with the notion of a threshold effect for creativity—high intelligence is necessary but not sufficient for creativity.

In our second study, we assessed cortical thickness in a cohort of sixty-one young adults, including the fifty-six from the spectroscopy study, using both measures of divergent thinking and creative achievement (Jung, Segall et al., 2010). We found cortical thickness in a region in the lingual gyrus was negatively associated with a psychometric measure of creative cognition, but was positively correlated with a different region in the right posterior cingulate. On measures of creative achievement, *less* gray matter volume in the left lateral orbitofrontal region was associated with higher creative achievement, but higher volume in the right angular gyrus correlated with creative achievement.

In our third study, we examined white matter integrity with a technique called diffusion tensor imaging (DTI), which measures the movement of water through myelinated axons. In a sample of seventy-two healthy young adults (including all of the previous subjects), we found an inverse relationship between white matter "integrity" (measured as "fractional anisotropy") and creative cognition in numerous regions within the left hemisphere, including the inferior frontal white matter and the superior longitudinal fasciculus (Jung, Grazioplene, Caprihan, Chavez, & Haier, 2010). The same relationship appeared in a small region within the right inferior frontal white matter and the anterior thalamic radiation. These three structural studies point to a decidedly left lateralized, frontosubcortical, and disinhibitory network of brain regions underlying creative cognition and achievement. These areas are summarized in figure 11.3 as part of the proposed F-DIM model of creativity. We describe this as a "model" as opposed to the P-FIT "theory" because it is based on a relatively few structural and lesion studies and is not readily testable until more experimental studies yield theoretical congruence.

These studies suggest to us that "less is more" with regard to creative cognition as measured by divergent thinking measures, particularly within frontosubcortical networks hypothesized to be central to creativity by several independent threads of thought (Dietrich, 2004; Flaherty, 2005; Heilman et al., 2003). The brain networks involved are likely *disinhibitory* in nature (Eysenck, 1995), with lesions and/or network degradation (i.e., cortical thinning, lower white matter coherence) located within a specific *network*, producing *increased* behavioral output. Central aspects of the network appear to include the frontal and temporal lobes, with cortical "tone" being modulated via interactions between the frontal lobes, basal ganglia and thalamus (part of the dopamine system) through white-matter pathways.

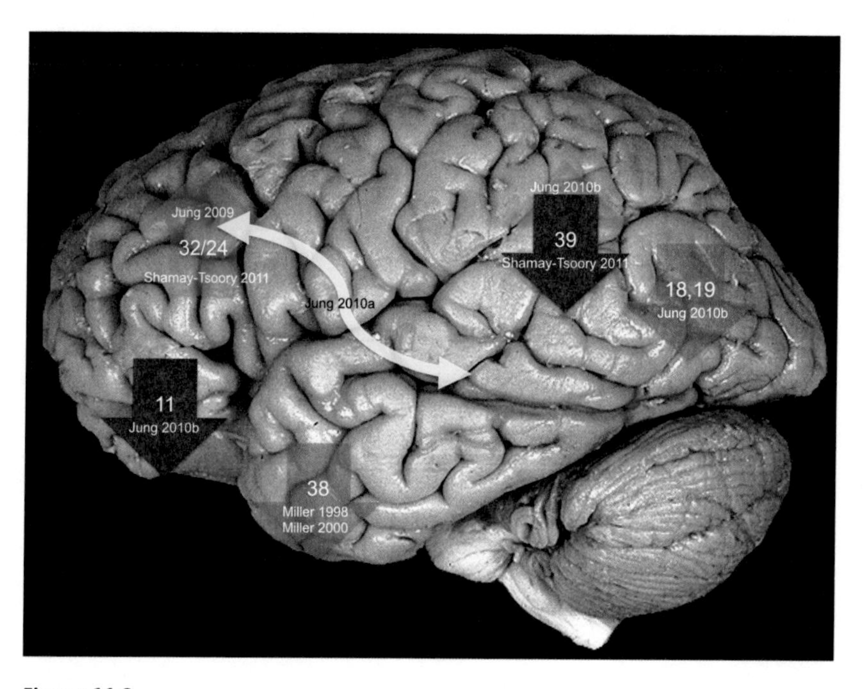

Figure 11.3
F-DIM model of creativity. Numbers indicate Brodmann areas. Blue = left lateralized; green = medial; purple = bilateral; yellow arrow = anterior thalamic radiation.

The story with intelligence may be similar—greater gray matter and better white-matter integrity go with higher scores but so does decreased function, which may reflect greater efficiency. That is to say that for both intelligence and creativity, we must look not only to *increased* neural tissue or activity in key brain regions (e.g., frontal lobes), but perhaps also to some mismatch between mutually excitatory and inhibitory brain regions (e.g., temporal lobes) that form a network subserving such complex human behaviors as creativity (e.g., planning, insight, inspiration). This notion of a delicate interplay of both *increases* and *decreases* in neural mass, white-matter organization, biochemical composition, and even functional activations within and between brain lobes and hemispheres is an important concept. Indeed, it is the rare brain that has highly developed networks of brain regions subserving intelligence (figure 11.2), and (concurrently) the somewhat underdeveloped network of brain regions associated with disinhibitory brain processes associated with creative cognition (figure 11.3). Such a finely tuned seesaw of complex higher and lower brain fidelity, balanced in dynamic opposition, would almost guarantee the rare occurrence of genius.

Why are we able to create a model for a network of creativity in the brain whereas three reviews (including our own) failed to do so? There are several reasons why the structural imaging and lesion studies might provide a more coherent model. First, the "lesion" method of cognitive neuroscience has long been considered the "gold standard" of methodology, showing the critical node necessary to a given cognitive task (Broca, 1861). Second, the "structural" techniques have all been demonstrated to have extremely high levels of reliability as measured with interclass correlations, with proton magnetic resonance spectroscopy being 0.98 (Gasparovic et al., in press), diffusion tensor imaging being 0.80 (Danielian, Iwata, Thomasson, & Floeter, 2010), and structural magnetic resonance imaging (sMRI) being 0.96 (Wonderlick et al., 2009); a major review of functional techniques (which predominated in the creativity reviews) revealed only modest reliability of 0.50 (Bennett & Miller, 2010). Similarly, our model focuses exclusively on measures of divergent thinking and achievement, such as the multiple uses test and creative achievement questionnaire, both of which have high reliability and validity (Domino & Domino, 2006). In contrast, many of the measures used in the creativity reviews were "home grown," extraordinarily diverse, and consisted of measures with unknown (and unknowable) reliability and/or validity, such as (1) composing a piece of music mentally (Petsche, 1996), (2) imagining a new design for a pen (Kowatari, et al., 2009), or (3) developing hypotheses about variations in quail eggs (Jin, Kwon, Jeong, Kwon, & Shin, 2006), to name a few. The use of standardized measures of divergent thinking (i.e., multiple uses test, Torrance Test of Creative Thinking), combined with lesion analysis and/or reliable imaging methodology (e.g., MRS, DTI, sMRI, and even fMRI), will help advance the field.

Brain Networks of Genius

Is There Brain Overlap for Intelligence and Creativity?

Given its rare and often idiosyncratic nature, it is not surprising that there is no systematic study of genius in the brain imaging literature. Our brain models of intelligence and creativity serve as a first approximation for identifying networks possibly related to genius. Particular regions within the network are customarily described in terms of Brodmann areas (BAs) in reference to Korbinian Brodmann, who first created a detailed cartography of the human brain in 1905 (Brodmann, 1905). Looking at figures

11.2 and 11.3, a qualitative analysis would suggest overlap between the P-FIT and F-DIM areas in four regions, including BAs 18/19 in the occipital lobe, BA 39 (the angular gyrus) in the parietal lobe, and BA 32 (the anterior cingulate gyrus) in the frontal lobe. Note that although there is overlap in these relatively large areas, only the anterior cingulate appears to show a consistent picture of higher fidelity associated with higher ability for both intelligence and creativity.

The anterior cingulate gyrus is a region of the brain ubiquitous in its involvement in numerous cognitive neuroscientific studies (Cabeza & Nyberg, 2000). However, the role of the anterior cingulate in intelligence, creativity, and genius might be more specific (Colom, Jung, & Haier, 2006; Frangou, Chitins, & Williams, 2004; Gong et al., 2005; Jung, Gasparovic, Chavez, Caprihan et al., 2009; Jung, Segall et al., 2010; Pfleiderer et al., 2004). The anterior cingulate cortex has been demonstrated to contain a unique type of spindle cell, found only in large hominoids, with double the frequency in humans than in great apes, suggesting strong selection pressures in this particular brain structure (Nimchinsky et al., 1999). The anterior cingulate gyrus appears to have some level of specificity with respect to the ability of our species to (1) down-regulate and/or activate broad networks of brain regions in service of divergent thinking, and subsequently (2) up-regulate and/or focus resources within frontal lobe networks in service of convergent thinking and/or persistence in pushing a new idea out into the world. Future studies will help further parse structure-function relationships within the anterior cingulate cortex if undertaken in subjects selected for high intelligence, or high creativity, or genius.

Where Are the Unique Areas of Nonoverlap between Figures 11.2 and 11.3?

Mostly, the areas shown in figures 11.2 and 11.3 do not overlap. Figure 11.2 represents a network of regions largely lateral (on the outer surface) and superior (toward the top half of the brain) in their distribution, whereas figure 11.3 represents a network of largely inferior (toward the lower half) and more medial (on the inner surfaces of the brain). For example, while intelligence was found to be associated with posterior brain regions including the extrastriate and fusiform gyri near the lateral occipital lobes, many creativity studies (including our structural studies) find associations with the cuneus and precuneus on the medial wall between the two hemispheres of the brain. Intelligence is associated with integrity of the dorsolateral prefrontal cortex: creative achievement with lower volumes of the orbitofrontal cortex, and increased creative drive in FTLD with damage to the left anterior temporal lobe. Intelligence is associated with integrity of white-matter tracts including the arcuate fasciculus and corpus callosum; divergent thinking and openness to experience were associated with lower measures of integrity within white-matter tracts linking the thalamus with frontal projection zones.

Intelligence and creativity appear to involve largely different brain networks. Tentatively, we interpret the former, focused on network integrity, to facilitate knowledge acquisition and retention, and the latter, focused on disinhibition of networks, to facilitate the generation of novel associations between knowledge stores. Whether there is a specific network for genius is not yet apparent.

The Mystery of the Einstein Area

One interesting anomaly when comparing figures 11.2 and 11.3 regards the inferior parietal lobule (BA 39). In studies of intelligence, this region has been implicated in better performance across studies, with greater volume, higher levels of the neuronal marker N-acetylaspartate, and greater functional activation all showing positive associations with measures of intelligence. In the "lesion" study of creativity reviewed above, however, subjects with damage to the left inferior parietal lobule performed better on measures of originality—some even performing better than normal control subjects. How can this be? Wasn't Einstein's brain unique by virtue of his inferior parietal lobule? But there is the rub: Einstein did not have a normal inferior parietal lobule, and the "abnormalities" were related to glial cells as opposed to neurons. Recall that Einstein had a 15 percent wider parietal lobe than controls and no parietal operculum (Witelson et al., 1999). However, subsequent studies showed that this greater parietal bulk was not comprised of neurons, but rather a higher number of glial cells—the support matrix of the brain (Diamond et al., 1985). Glial cells have customarily been viewed as the "glue" (literally what "glia" translates to) that holds the brain together. However, glial cells comprise 85 percent of the total brain volume, and we are entering into an era where the neuron doctrine of brain function is being slowly adapted to include the "glial doctrine," with recent studies demonstrating long-range communication between glial populations and glial modulation of neuronal tone (Fields & Stevens-Graham, 2002). Thus, Einstein's brain is not entirely inconsistent with a "lesion" model of genius comprising a network of brain regions, some with greater neuronal fidelity, some with lowered (i.e., disinhibitory) characteristics.

Conclusion

Complex phenomena like intelligence, creativity, and genius can be studied scientifically with modern neuroscience methods even as their definitions evolve with better empirical observations. Indeed, very clever individuals, such as Hans Eysenck, formulated hypotheses regarding genius without the benefit of sophisticated neuroimaging techniques that we now take for granted. These hypotheses were not too far off the mark given the benefit of careful interrogation with the tools and techniques of the modern neuroscientist. This has always been the case in science, whether investigating the nature of an atom or a gene or a memory. Such investigations typically raise more questions than they resolve; however, asking the right questions is key. Will an understanding of the neural basis of intelligence or creativity, or even genius, change how we approach education? There are already moves in this direction based on very tentative data (Ramsden et al., 2011) but much more research is necessary. As always, caution is required, but the future of creativity research looks bright indeed.

References

Abraham, C. (2002). *Possessing genius: The bizarre odyssey of Einstein's brain*. New York: St. Martin's Press.

Amabile, T. M. (1982). Social psychology of creativity: A consensual assessment technique. *Journal of Personality and Social Psychology, 43*(5), 997–1013.

Anderson, B., & Harvey, T. (1996). Alterations in cortical thickness and neuronal density in the frontal cortex of Albert Einstein. *Neuroscience Letters, 210*(3), 161–164.

Arden, R., Chavez, R. S., Grazioplene, R., & Jung, R. E. (2010). Neuroimaging creativity: A psychometric view. *Behavioural Brain Research, 214*(2), 143–156.

Bennett, C. M., & Miller, M. B. (2010). How reliable are the results from functional magnetic resonance imaging? *Annals of the New York Academy of Sciences, 1191*, 133–155.

Bishop, S. J., Fossella, J., Croucher, C. J., & Duncan, J. (2008). COMT val(158)met genotype affects recruitment of neural mechanisms supporting fluid intelligence. *Cerebral Cortex, 18*(9), 2132–2140.

Brain of Einstein continues peregrinations. (1981). *Science*, 213, 521.

Brain that rocked physics rests in cider box. (1978). *Science*, 201, 696.

Brian, D. (1996). *Einstein: A life*. New York: Wiley & Sons.

Broca, M. P. (1861). Remarques sur le siege de la faculte du langage articule suivies d'une observation d'aphemie. *Bulletins de la Société Anatomique de Paris, 36,* 330–357.

Brodmann, K. (1905). Beiträge zur histologischen Lokalisation der Grosshirnrinde: dritte Mitteilung: Die Rindenfelder der niederen Affen. *Journal für Psychologie und Neurologie, 4,* 177–226.

Cabeza, R., & Nyberg, L. (2000). Imaging cognition II: An empirical review of 275 PET and fMRI studies. *Journal of Cognitive Neuroscience, 12*(1), 1–47.

Chiang, M. C., Barysheva, M., Shattuck, D. W., Lee, A. D., Madsen, S. K., Avedissian, C., et al. (2009). Genetics of brain fiber architecture and intellectual performance. *Journal of Neuroscience, 29*(7), 2212–2224.

Colom, R., Haier, R. J., Head, K., Alvarez-Linera, J., Quiroga, M. A., Shih, P. C., et al. (2009). Gray matter correlates of fluid, crystallized, and spatial intelligence: Testing the P-FIT model. *Intelligence, 37*(2), 124–135.

Colom, R., Jung, R. E., & Haier, R. J. (2006). Distributed brain sites for the G-factor of intelligence. *NeuroImage, 31*(3), 1359–1365.

Costa, P. T., & McCrae, R. R. (1992). *NEO PI-R professional manual*. Odessa, FL: Psychological Assessment Resources.

Crammond, D. J. (1997). Motor imagery: Never in your wildest dreams. *Trends in Neurosciences, 20,* 54–57.

Danielian, L. E., Iwata, N. K., Thomasson, D. M., & Floeter, M. K. (2010). Reliability of fiber tracking measurements in diffusion tensor imaging for longitudinal study. *NeuroImage, 49*(2), 1572–1580.

Deary, I. J., Penke, L., & Johnson, W. (2010). The neuroscience of human intelligence differences. *Nature Reviews: Neuroscience, 11*(3), 201–211.

Diamond, M. C., Scheibel, A. B., Murphy, G. M., Jr., & Harvey, T. (1985). On the brain of a scientist: Albert Einstein. *Experimental Neurology, 88*(1), 198–204.

Dietrich, A. (2004). The cognitive neuroscience of creativity. *Psychonomic Bulletin & Review, 11*(6), 1011–1026.

Dietrich, A., & Kanso, R. (2010). A review of EEG, ERP, and neuroimaging studies of creativity and insight. *Psychological Bulletin, 136*(5), 822–848.

Domino, G., & Domino, M. L. (2006). *Psychological testing* (2nd Ed.). New York: Cambridge University Press.

Einstein, A. (1905a). Ist die Trägheit eines Körpers von seinem Energiegehalt abhängig? *Annalen der Physik, 18,* 639–641.

Einstein, A. (1905b). Uber die von der molekularkinetischen theorie der wärme geforderte bewegung von in ruhenden flüssigkeiten suspendierten teilchen. *Annalen der Physik, 17,* 549–560.

Einstein, A. (1905c). Uber einen die Erzeugung und Verwandlung des Lichtes betreffenden heuristischen Gesichtpunkt. *Annalen der Physik, 17,* 132–148.

Einstein, A. (1905d). Zur Elektrodynamik bewegter Körper. *Annalen der Physik, 17,* 891–921.

Eysenck, H. (1995). *Genius: The natural history of creativity.* Cambridge: Cambridge University Press.

Falk, D. (2009). New information about Albert Einstein's brain. *Frontiers in Evolutionary Neuroscience, 1*(3). doi: 10.3389/neuro.18.003.2009.

Fields, R. D., & Stevens-Graham, B. (2002). New insights into neuron-glia communication. *Science, 298*(5593), 556–562.

Flaherty, A. W. (2011). Brain illness and creativity: Mechanisms and treatment risks. *Canadian Journal of Psychiatry, 56*(3), 132–143.

Flaherty, A. W. (2005). Frontotemporal and dopaminergic control of idea generation and creative drive. *Journal of Comparative Neurology, 493*(1), 147–153.

Frangou, S., Chitins, X., & Williams, S. C. (2004). Mapping IQ and gray matter density in healthy young people. *NeuroImage, 23*(3), 800–805.

Galaburda, A. M. (1999). Albert Einstein's brain. *Lancet, 354,* 1821.

Gasparovic, C., Bedrick, E., Mayer, A. R., Yeo, R. A., Calhoun, V. C., & Jung, R. E. (in press). Test-retest reliability of short-echo-time spectroscopic imaging data from human brain at 3T. *Magnetic Resonance in Medicine.*

Gong, Q. Y., Sluming, V., Mayes, A., Keller, S., Barrick, T., Cezayirli, E., et al. (2005). Voxel-based morphometry and stereology provide convergent evidence of the importance of medial prefrontal cortex for fluid intelligence in healthy adults. *NeuroImage, 25*(4), 1175–1186.

Haier, R. J. (2009). Neuro-intelligence, neuro-metrics, and the next phase of brain imaging studies. *Intelligence, 37*(2), 121–123.

Harlow, J. M. (1848). Passage of an iron rod through the head. *Boston Medical and Surgical Journal, 39,* 389–393.

Heilman, K. M., Nadeau, S. E., & Beversdorf, D. O. (2003). Creative innovation: Possible brain mechanisms. *Neurocase, 9*(5), 369–379.

Highfield, R., & Carter, P. (1993). *The private lives of Albert Einstein.* New York: St Martin's Press.

Hines, T. (1998). Further on Einstein's brain. *Experimental Neurology, 150,* 343–344.

Jensen, A. R. (1998). *The G factor: The science of mental ability.* New York: Praeger.

Jin, S. H., Kwon, Y. J., Jeong, J. S., Kwon, S. W., & Shin, D. H. (2006). Differences in brain information transmission between gifted and normal children during scientific hypothesis generation. *Brain and Cognition, 62*(3), 191–197.

Jung, R. E., Gasparovic, C., Chavez, R. S., Caprihan, A., Barrow, R., & Yeo, R. A. (2009). Imaging intelligence with proton magnetic resonance spectroscopy. *Intelligence, 37*(2), 192–198.

Jung, R. E., Gasparovic, C., Chavez, R. S., Flores, R. A., Smith, S. M., Caprihan, A., et al. (2009). Biochemical support for the "threshold" theory of creativity: A magnetic resonance spectroscopy study. *Journal of Neuroscience, 29*(16), 5319–5325.

Jung, R. E., Grazioplene, R., Caprihan, A., Chavez, R. S., & Haier, R. J. (2010). White matter integrity, creativity, and psychopathology: disentangling constructs with diffusion tensor imaging. *PLoS ONE, 5*(3), e9818.

Jung, R. E., & Haier, R. J. (2007). The Parieto-Frontal Integration Theory (P-FIT) of intelligence: Converging neuroimaging evidence. *Behavioral and Brain Sciences, 30,* 135–154.

Jung, R. E., Segall, J. M., Jeremy Bockholt, H., Flores, R. A., Smith, S. M., Chavez, R. S., et al. (2010). Neuroanatomy of creativity. *Human Brain Mapping, 31*(3), 398–409.

Karama, S., Ad-Dab'bagh, Y., Haier, R. J., Deary, I. J., Lyttelton, O. C., Lepage, C., et al. (2009). Positive association between cognitive ability and cortical thickness in a representative US sample of healthy 6 to 18 year-olds. *Intelligence, 37*(4), 431–442.

Kowatari, Y., Lee, S. H., Yamamura, H., Nagamori, Y., Levy, P., Yamane, S., et al. (2009). Neural networks involved in artistic creativity. *Human Brain Mapping, 30*(5), 1678–1690.

Li, Y. H., Liu, Y., Li, J., Qin, W., Li, K. C., Yu, C. S., et al. (2009). Brain anatomical network and intelligence. *PLoS Computational Biology, 5*(5), 17.

Liu, B., Li, J., Yu, C. S., Li, Y. H., Liu, Y., Song, M., et al. (2010). Haplotypes of catechol-O-methyltransferase modulate intelligence-related brain white matter integrity. *NeuroImage, 50*(1), 243–249.

Luders, E., Narr, K. I., Bilder, R. M., Szeszko, P. R., Gurbani, M. N., Hamilton, L., et al. (2008). Mapping the relationship between cortical convolution and intelligence: Effects of gender. *Cerebral Cortex, 18*(9), 2019–2026.

Luders, E., Thompson, P. M., Narr, K. L., Zamanyan, A., Chou, Y. Y., Gutman, B., et al. (2011). The link between callosal thickness and intelligence in healthy children and adolescents. *NeuroImage, 54*(3), 1823–1830.

Lythgoe, M. F., Pollak, T. A., Kalmus, M., de Haan, M., & Chong, W. K. (2005). Obsessive, prolific artistic output following subarachnoid hemorrhage. *Neurology, 64*(2), 397–398.

Mendez, M. F. (2005). Hypergraphia for poetry in an epileptic patient. *Journal of Neuropsychiatry and Clinical Neurosciences, 17*, 560–561.

Menzel, C., Grunwald, F., Klemm, E., Ruhlmann, J., Elger, C. E., & Biersack, H. J. (1998). Inhibitory effects of mesial temporal partial seizures onto frontal neocortical structures. *Acta Neurologica Belgica, 98*(4), 327–331.

Miller, B. L., Boone, K., Cummings, J. L., Read, S. L., & Mishkin, F. (2000). Functional correlates of musical and visual ability in frontotemporal dementia. *British Journal of Psychiatry, 176*, 458–463.

Miller, B. L., Cummings, J., Mishkin, F., Boone, K., Prince, F., Ponton, M., et al. (1998). Emergence of artistic talent in frontotemporal dementia. *Neurology, 51*(4), 978–982.

Mink, J. W. (1996). The basal ganglia: Focused selection and inhibition of competing motor programs. *Progress in Neurobiology, 50*(4), 381–425.

Nimchinsky, E. A., Gilissen, E., Allman, J. M., Perl, D. P., Erwin, J. M., & Hof, P. R. (1999). A neuronal morphologic type unique to humans and great apes. *Proceedings of the National Academy of Sciences of the United States of America, 96*(9), 5268–5273.

Paterniti, M. (2000). *Driving Mr. Albert: A trip across America with Einstein's brain*. New York: Dial Press.

Petsche, H. (1996). Approaches to verbal, visual, and musical creativity by EEG coherence analysis. *International Journal of Psychophysiology, 24*(1–2), 145–159.

Pfleiderer, B., Ohrmann, P., Suslow, T., Wolgast, M., Gerlach, A. L., Heindel, W., et al. (2004). N-acetylaspartate levels of left frontal cortex are associated with verbal intelligence in women but not in men: A proton magnetic resonance spectroscopy study. *Neuroscience, 123*(4), 1053–1058.

Ramsden, S., Richardson, F. M., Josse, G., Thomas, M. S., Ellis, C., Shakeshaft, C., et al. (2011). Verbal and non-verbal intelligence changes in the teenage brain. *Nature, 479*, 113–116.

Rankin, K. P., Liu, A. L. A., Howard, S., Slama, H., Hou, C. E., Shuster, K., et al. (2007). A case-controlled study of altered visual art production in Alzheimer's and FTLD. *Cognitive and Behavioral Neurology, 20*(1), 48–61.

Ro, T., Farne, A., Johnson, R. M., Wedeen, V., Chu, Z., Wang, Z. J., et al. (2007). Feeling sounds after a thalamic lesion. *Annals of Neurology, 62*(5), 433–441.

Runco, M. A. (2004). Creativity. *Annual Review of Psychology, 55*, 657–687.

Runco, M. A., & Charles, R. E. (1993). Judgments of originality and appropriateness as predictors of creativity. *Personality and Individual Differences, 15*(5), 537–546.

Sawyer, K. (2011). The cognitive neuroscience of creativity: A critical review. *Creativity Research Journal, 23*(2), 137–154.

Schmithorst, V. J. (2009). Developmental sex differences in the relation of neuroanatomical connectivity to intelligence. *Intelligence, 37*(2), 164–173.

Schrag, A., & Trimble, M. (2001). Poetic talent unmasked by treatment of Parkinson's disease. *Movement Disorders, 16*, 1175–1176.

Scoville, W. B., & Milner, B. (1957). Loss of recent memory after bilateral hippocampal lesions. *Journal of Neurology, Neurosurgery, and Psychiatry, 20*(1), 11–21.

Shamay-Tsoory, S. G., Adler, N., Aharon-Peretz, J., Perry, D., & Mayseless, N. (2011). The origins of originality: The neural bases of creative thinking and originality. *Neuropsychologia, 29*, 178–185.

Snyder, A. W., & Mitchell, D. J. (1999). Is integer arithmetic fundamental to mental processing? The mind's secret arithmetic. *Proceedings of the Royal Society of London, Series B: Biological Sciences, 266*(1419), 587–592.

Spitzka, E. A. (1907). A study of the brains of six eminent scientists and scholars belonging to the American Anthropometric Society. *Transactions of the American Philosophical Society, 21*, 175–308.

Tamnes, C. K., Ostby, Y., Walhovd, K. B., Westlye, L. T., Due-Tonnessen, P., & Fjell, A. M. (2010). Intellectual abilities and white matter microstructure in development: A diffusion tensor imaging study. *Human Brain Mapping, 31*(10), 1609–1625.

Tang, C. Y., Eaves, E. L., Ng, J. C., Carpenter, D. M., Mai, X., Schroeder, D. H., et al. (2010). Brain networks for working memory and factors of intelligence assessed in males and females with fMRI and DTI. *Intelligence, 38*(3), 293–303.

Thomas-Anterion, C., Creac'h, C., Dionet, E., Borg, C., Extier, C., Faillenot, I., & Peyron, R. (2010). De novo artistic activity following insular-SII ischemia. *Pain, 150*(1), 121–127.

van den Heuvel, M. P., Stam, C. J., Kahn, R. S., & Pol, H. E. H. (2009). Efficiency of functional brain networks and intellectual performance. *Journal of Neuroscience, 29*(23), 7619–7624.

Witelson, S. F., Kigar, D. L., & Harvey, T. (1999). The exceptional brain of Albert Einstein. *Lancet, 353*(9170), 2149–2153.

Wonderlick, J. S., Ziegler, D. A., Hosseini-Varnamkhasti, P., Locascio, J. J., Bakkour, A., van der Kouwe, A., et al. (2009). Reliability of MRI-derived cortical and subcortical morphometric measures: Effects of pulse sequence, voxel geometry, and parallel imaging. *NeuroImage, 44*(4), 1324–1333.

VI Aesthetic and Creative Products

12 Fostering Creativity: Insights from Neuroscience

Oshin Vartanian

There is now general consensus that creativity is a componential trait. As such, it stands to reason that fostering any of its underlying components should in turn benefit creative production. With some variation, the three fundamental components of creativity are considered to be *motivation* (task commitment), *ability* (domain expertise), and *creative thinking skills* (Amabile, 1998; Renzulli, 1986). Accordingly, a large body of behavioral evidence has demonstrated that creativity is enhanced when motivation is intrinsically oriented (Amabile, 1985), levels of ability and expertise are high (Weisberg, 1999), and creative thinking skills (e.g., divergent thinking) are superior (Plucker & Renzulli, 1999).

However, can evidence gleaned from the structure and function of the brain enhance our ability to foster creativity? I argue that it can, for two reasons. First, interventions designed to enhance motivation, abilities, and skills must be realized in the brain and therefore have traceable neural correlates, which in turn can be used to verify that learning has occurred. In fact, much evidence now points to training-based neural plasticity in the brain—both structurally and functionally. For example, Schlaug et al. (2009) have demonstrated structural changes in the anterior midbody of the corpus callosum (which connects the premotor and supplementary motor areas of the two hemispheres) following musical training. Similarly, functional changes in the brain's frontoparietal system have been shown to occur following training of working memory (WM) (see Klingberg, 2010)—a critical ability for most types of creativity. These findings suggest that researchers can use data derived from brain structure and function to confirm and in turn optimize training regimens for creativity.

A related but conceptually separable contribution the neurosciences can make to foster creativity is the information they provide on the brain's metabolic rate. Specifically, the uptake of glucose in positron emission tomography (PET) and the blood oxygen-level dependent (BOLD) signal in functional magnetic resonance imaging (fMRI) provide direct and indirect information about the metabolic requirements of various cognitive tasks. Interventions (e.g., cognitive training) that increase the brain's neural efficiency (by reducing its metabolic demands) for creative tasks provide another avenue for facilitating creativity. For example, it is now known that, contrary to conventional wisdom, the levels and stability of glucose in the brain vary such that it is not always present in ample amounts to optimally support learning and memory (see McNay, McCarty, & Gold, 2001). Thus, given the resource limitations that exist in the brain, interventions that lower the limiting metabolic thresholds for any component of creative cognition should theoretically facilitate its occurrence.

In this chapter, I review two strands of neuroscientific research that I believe have shown promise in fostering creativity. The first involves enhancement of WM and fluid intelligence through cognitive training. Given that WM ability and fluid intelligence are related positively to many forms of creativity (Nusbaum & Silvia, 2011; Sligh, Conners, & Roskos-Ewoldsen, 2005), the ability to use neuroscientific data to track enhancement of WM and fluid intelligence in the brain can be used to optimize creativity training. In this sense, neuroscientific data can play a confirmatory role in establishing training effects. The second strand of relevant research is focused on the concept of neural efficiency and on the utility of cognitive interventions to increase neural efficiency for creativity tasks. Although this literature is relatively limited compared to the literature on the enhancement of WM and fluid intelligence, emerging evidence suggests that it is possible to make the neural processing more efficient through behavioral intervention (see Vartanian et al., 2013).

Improving Working Memory

Working memory involves the ability to maintain and manipulate information. WM is hypothesized to play an important role in most types of creativity. For example, one of the most common engines for the generation of creative ideas is the novel and useful combination of concepts previously thought to be unrelated (Poincaré, 1913; Vartanian, Martindale, & Matthews, 2009; see also Sternberg, 1999). It is immediately evident that this combinatorial engine requires the maintenance and manipulation of two or more concepts—for which WM is a necessity. Thus, the question arises: can we improve WM capacity through training? The answer, based on evidence from a large number of studies that have investigated the impact of cognitive or brain training on cognitive capacity—specifically WM—appears to be yes (for reviews see Klingberg, 2010; Morrison &

Chein, 2011). The central question in this research area is not whether performance on any given cognitive task can be improved by training—it has long been known that it can—but whether improvements in a cognitive task as a function of WM training can *transfer* to other untrained tasks or to improvements in general cognitive function. The issue of transfer is critical for fostering creativity because ideally it would be possible to train on a small set of "core" tasks and observe improvements in a variety of target creativity tasks.

On balance, the evidence suggests that there is reason to be optimistic about transfer effects as a function of WM training, although a number of qualifiers apply. First, for the most part studies that have shown transfer effects have been conducted within laboratory settings in which a strict regimen of training has been administered by trained personnel. Transfer effects have not been shown for individuals outside of the lab engaged in unsupervised daily cognitive training (Owen et al., 2010). Second, training effects are more likely to transfer if subjects train on "core" WM tasks, likely because such training targets domain-general abilities that underlie many different target activities. In contrast, although "strategy training" can result in improvements in the specific task used for training, it tends not to exhibit far-reaching transfer effects. Finally, although training and transfer effects have been demonstrated in young adults, the evidence is less convincing for seniors. For example, some data with older adults suggest that the improvements related to cognitive training extend only to the trained task and sometimes to closely related memory measures (Li et al., 2008). However, when researchers have used ecologically valid measures of verbal learning and "everyday attention," as well as self-reported functional measures, transfer of improvements has been observed (Richmond et al., 2011). This suggests that cognitive training could be efficacious in older adults if transfer effects were studied using ecologically valid target tasks that more closely resemble the real-life activities of older adults. However, there is also evidence to suggest that in this cohort (compared to young adults) glucose may be depleted much more quickly from key brain regions during cognitive performance (McNay et al., 2001), such that effective training may have to be coupled with external metabolic interventions.

The conclusion from this large body of literature is that given the right conditions, WM training does transfer to improvements in target tasks as well as general cognitive function. To the extent that creativity relies on WM, this finding is very promising for interventions aimed at improving the "ability" component of creative production (Renzulli, 1986).

What is the impact, if any, of WM training on neural function? The answer to this question is rather nuanced. Recently, Klingberg (2010) reviewed ten fMRI studies that had investigated the impact of repeated performance of WM tasks on neural function. These studies could be divided into two categories. The first category involved studies ($N = 6$) in which researchers investigated the impact of repeated training on the *trained* tasks themselves. Five of these studies were structurally similar in that the impact of repeated WM training was examined based on relatively short training periods ranging from thirty minutes to two hours. The training tasks included delayed matching-to-sample and object, verbal, and spatial WM tasks. The pattern of results across studies was surprisingly consistent, uniformly demonstrating reductions in the BOLD response in various structures including the precentral sulcus, occipital lobe, parietal lobe, cingulate cortex, dorsal lateral prefrontal cortex (DLPFC), frontopolar cortex, and inferior frontal cortex. One of the six studies that used an entirely different design generated a different pattern of results. Specifically, Hempel et al. (2004) instructed their participants to train on the visuospatial n-back daily for four weeks. Briefly, on each trial of the n-back task the participant must decide whether the information currently present matches information presented a specified number of trials earlier—whether auditory or visual. By definition, optimal performance necessitates updating the contents of WM as a function of instructions. Hempel et al. (2004) tested the effect of training on neural function twice—two weeks and four weeks after training. Whereas activation in DLPFC and parietal lobe was elevated at two weeks, it was reduced at four weeks. Thus, aside from Hempel et al.'s (2004) study, the overall pattern of results suggests that when training periods are relatively short and the target is the trained task, reductions in the BOLD signal will be observed. This pattern closely resembles early PET studies in which repeated practice with effortful tasks reduced the extent of brain activation over time (for review see Jung & Haier, 2007).

The second category involved studies ($N = 4$) in which researchers investigated the impact of repeated training on transfer to other *untrained* tasks. Furthermore, these studies relied on longer and more frequent training regimens, ranging from ten hours over ten days to twenty hours over ten weeks. The tasks included updating, and verbal, object, and visuospatial WM tasks. Compared to training studies in which the effects were examined on the trained tasks, these studies generated a rather heterogeneous pattern of results including increases in the BOLD response in DLPFC, parietal lobe, caudate, and left inferior frontal cortex, accompanied by decreases in the BOLD response in DLPFC, parietal lobe, and the cingulate. It is important to note that despite differences in the direction of the effect across these four studies, the effects were localized in the WM frontoparietal network and the basal ganglia, a structure involved in selection of relevant information in WM tasks (McNab & Klingberg, 2008).

In conclusion, when combined together, the results of fMRI training studies based on trained and transferred skills are consistent with the interpretation that training targets the network of brain regions implicated in domain-general aspects of WM including DLPFC, parietal cortex, and basal ganglia (Wager & Smith, 2003). Furthermore, there is reason to believe that transfer from the trained task to the target task will be facilitated to the extent that the two tasks recruit overlapping cortical regions (see Morrison & Chein, 2011). Despite these encouraging results, a number of key issues remain unresolved. First, because activation in the frontoparietal network is also correlated with task difficulty (Barch et al., 1997), at the moment it is not possible to determine with certainty whether its responsiveness to training is a function of WM engagement or whether it is a by-product of responsiveness to task difficulty. For example, the reduced BOLD response in the frontoparietal network following short WM training regimens may simply be because the task becomes less difficult over time, and not necessarily because of improved WM capacity. Second, repeated training on WM tasks may involve changes to factors other than WM capacity, such as priming, automaticity, familiarity, and strategy learning. This means that although, based on improvements in performance as well as targeted variation in neural function, repeated WM training appears to "work" (Morrison & Chein, 2011), it is more difficult to isolate specifically which of the underlying factors associated with repeated training is responsible for the observed effects (see also Buschkuehl, Jaeggi, & Jonides, 2012).

Improving Fluid Intelligence

Fluid intelligence is defined as the ability to adapt to new situations and is characterized by increased abstraction and complexity in thinking (Cattell, 1963). In contrast, crystallized intelligence is defined as repository of knowledge and skills acquired through learning and experience. Not surprisingly, crystallized intelligence tends to increase across the lifespan as people acquire new knowledge and learn new skills (Deary, Penke, & Johnson, 2010). Furthermore, it has long been known that it is possible to boost crystallized intelligence through teaching. In contrast, fluid intelligence has historically not been viewed as an improvable ability. Coupled with the fact that it has a very high heritability quotient, fluid intelligence has historically been considered to be strongly influenced by genetics, although recent evidence suggests that plasticity is retained at least into the teenage years (Ramsden et al., 2011).

This traditional view was challenged by an important recent study conducted by researchers at the University of Michigan who reported evidence demonstrating that it was in fact possible to boost fluid intelligence. Specifically, Jaeggi, Buschkuehl, Jonides, and Perrig (2008) administered a dual n-back task to participants in the experimental condition, who trained for various number of sessions ranging among 8, 12, 17, and 19 days. The dual n-back task used was very effortful as it required simultaneous tracking of visual and auditory information. In addition, its difficulty level was adjusted in relation to individual performance. To assess fluid intelligence, participants were assessed using either Raven's Advanced Progressive Matrices (RAPM) test or the short version of the Bochumer Matrizen-Test (BOMAT) prior to and following the completion of training. The results demonstrated that WM capacity (as measured by mean n-back level) increased over the course of training, and that longer training resulted in greater gain. In addition, compared to a passive control condition, participants in the experimental condition demonstrated significant gain in fluid intelligence based on pre-post-change in test scores. Perhaps most interestingly, there was a dose-response effect such that a longer frequency of training was correlated with greater gain in fluid intelligence.

The same team subsequently tested the effectiveness of cognitive training in elementary and middle school children by means of a videogame-like WM task (Jaeggi, Buschkuehl, Jonides, & Shah, 2011). The results largely supported earlier findings, with the caveat that gains in fluid intelligence were observed only in those children who exhibited significant WM improvement in the course of training. This is perhaps not surprising, given that variation in the motivation to perform may play a bigger role in children than it does in adults and would therefore be an important consideration. Regardless, the set of studies by Jaeggi and colleagues demonstrates that it is possible to boost fluid intelligence, and that the gains based on WM capacity are transferrable to fluid intelligence. This feature opens up the possibility to boost performance in target tasks that rely on fluid intelligence through WM training, potentially opening up a number of applications in educational and professional domains (Sternberg, 2008).

Fluid Intelligence and Creativity

Given the aforementioned definition of fluid intelligence, it is rather surprising that this construct has not played a more prominent role in studies of creativity. Perhaps some of this may have to do with the fact that psychometric intelligence and creativity have shown a weak relationship across studies in the past. For example, regarding the correlation between intelligence and creativity, a recent meta-analysis of 447 effect sizes reported an average weighted effect size of only r = 0.174 (Kim, 2005). However, more recent evidence suggests that creativity loads heavily on *fluid* intelligence. For example, Nusbaum and Silvia (2011) used modern approaches to creativity assessment and latent variable modeling to investigate the role of fluid intelligence in divergent thinking. They reported two important findings. First, the effect of fluid intelligence on divergent thinking was mediated by executive switching (study 1). Executive switching was measured as a function of the number of times subjects switched idea categories during divergent thinking. Subsequently, the researchers investigated the extent to which participants were able to implement an effective strategy for an unusual uses task (study 2). The results demonstrated that only those participants high in fluid intelligence were able to do so, consistent with their higher ability to maintain access to the strategy and use it despite interference. The results of Nusbaum and Silvia (2011) demonstrate that fluid intelligence underlies the extent to which participants can implement the tools necessary for divergent thinking, including category shifts and strategy use (see also Sligh et al., 2005).

Very similar conclusions can be drawn from the work of Gilhooly, Fioratou, Anthony, and Wynn (2007), who used a think-aloud strategy to analyze the characteristics of the responses that were generated in an alternate uses task. They demonstrated that earlier uses generated in response to a prompt are frequently not creative, and involve well-known uses derived from long-term memory. Having exhausted mnemonic recall, uses generated later in the sequence are frequently more creative and involve strategies that load on executive function, such as disassembly and reassembly of parts (experiment 1). Subsequently, they showed that the generation of new creative uses was predicted by performance on letter fluency—an executive loading task (experiment 2). The results of Gilhooly et al. (2007) are congruent with those reported by Nusbaum and Silvia (2011), suggesting that despite the low correlations reported between creativity and intelligence in earlier studies (see Kim, 2005), creativity does load on fluid intelligence and executive function. Of course, given Jaeggi et al.'s (2008, 2011) demonstrations that fluid intelligence can be improved by a regimen of repeated training on the n-back task, this brings up the obvious question: can creativity be improved by improving WM capacity based on training on the n-back task? Before we answer this question, we will take a short detour and briefly review the current state of knowledge about the neuroscience of intelligence (for a more in-depth review see Jung & Haier, this vol.).

Neuroscience of Intelligence

The neuroscience of intelligence has been the focus of much interest recently (for reviews see Deary et al., 2010; Jung & Haier, 2007, this vol.). Based on a large-scale review of all available structural and functional neuroimaging studies, Jung and Haier (2007) proposed the parieto-frontal integration theory of intelligence (P-FIT). This theory localizes individual differences in intelligence to specific frontal and parietal regions (see Jung & Haier, 2007; figure 11.2, this vol.). There are two issues of relevance for the purposes of the current chapter that must be highlighted. First, the neural correlates of general *fluid* intelligence are embedded within P-FIT, in particular in the lateral prefrontal cortex (Duncan et al., 2000; Gray, Chabris, & Braver, 2003). Interestingly, these are the same brain regions that are heavily implicated in inhibition and (executive) control of attention (Aron, Robbins, & Poldrack, 2004). This is not surprising, given that an important component of intelligent behavior involves selective attention to relevant information.

Second, although by no means a universal observation, a large number of studies have reported an inverse relation between fluid intelligence and metabolic rate in a variety of cognitive tasks (Deary et al., 2010; Jung & Haier, 2007, this vol.; Fink & Neubauer, this vol.; Neubauer, Fink, & Schrausser, 2002; Neubauer, Grabner, Fink, & Neuper, 2005). Furthermore, a recent review suggested that this inverse relation is most likely to be observed in the frontal cortex (Neubauer & Fink, 2009). This inverse relation has been interpreted to mean that intelligence involves a more efficient use of cortical resources during task performance—and is therefore referred to as the *neural efficiency hypothesis*. These findings suggest that gains in fluid intelligence as a function of WM training should be localizable in the prefrontal cortex and that they may correlate inversely with brain activation during performance on tasks that recruit fluid intelligence.

Working Memory, Fluid Intelligence, and Creativity

Given that WM training can boost fluid intelligence, coupled with the observation that variations in fluid intelligence are correlated with activation in the prefrontal cortex, Vartanian et al. (2013) recently conducted a study to test the hypothesis that WM training could be used to enhance neural efficiency in the prefrontal cortex in the course of creative problem solving. For creative problem solving, they opted to use a divergent thinking paradigm. This decision was made for three reasons. First, it is known that divergent thinking loads heavily on fluid intelligence (Nusbaum & Silvia, 2011), making it likely that boosting fluid intelligence by WM training will be advantageous for divergent production. Second, previous fMRI and neuropsychological studies of divergent thinking have pinpointed the lateral prefrontal cortex as a key region for solution generation (Goel & Vartanian, 2005; Miller & Tippett, 1996). Recall that this same region is viewed as a major hub for fluid intelligence in the brain (Duncan et al., 2000; Gray et al., 2003). Finally, Fink et al. (2009) had developed an experimental design for administering the alternate uses task (AUT) in the fMRI scanner that could be reemployed in Vartanian et al.'s (2013) study, with minor modifications.

Furthermore, based on recommendations offered in previous reviews of the cognitive (brain) training literature (Klingberg, 2010; Morrison & Chein, 2011), our laboratory implemented certain methodological features. First, whereas the experimental group trained on the n-back task, we employed an active (rather than passive) control group that trained on an easy four-choice reaction-time (RT) task not expected to load on WM. Second, the participants in the experimental ($N = 17$) and control ($N = 17$) conditions were matched for sex and age. Finally, the target task used in the scanner was different from the task used for training, enabling a test of transfer effects.

In the beginning of the experiment all participants were tested on one of two versions (odd or even) of RAPM (see Jaeggi et al., 2008). There was no difference in average baseline RAPM scores between the experimental and control groups. Then, whereas the experimental group trained on three separate days on the 2-back and 3-back tasks (in alternating blocks within the same session), the control group trained on three separate days on an easy four-choice RT task. The results demonstrated that, for the experimental group, performance improved across the three training sessions. We attributed this improvement to a gain in WM capacity. In contrast, and as predicted, performance was at ceiling across all three sessions for the control group. Following training, all participants were tested on the version of RAPM not administered to them at baseline. We computed post-pre RAPM scores to calculate gain in fluid intelligence. The results demonstrated that whereas there was significant gain in fluid intelligence in the experimental group, no gain was observed in the control group. In

Figure 12.1
Task structure for alternate uses task (Vartanian et al., 2013). ms = milliseconds. Participants were instructed to generate uses of objects (see text).

summary, our results demonstrated that repeated training on the n-back task boosted (a) WM capacity and (b) fluid intelligence.

Next, to investigate the transfer of gains in fluid intelligence to divergent thinking, we administered the AUT in the fMRI scanner. The AUT was modeled after Fink et al. (2009) (figure 12.1). The task was presented in two blocks. In the *uses* block participants were presented with names of common objects (e.g., fork) and instructed to think of as many uses for them as possible. In the *characteristics* block participants were instructed not to generate uses, but to recall physical features characteristic of the presented object instead.

Here the focus will be on the most relevant behavioral and fMRI findings. First, contrary to our prediction, gains in WM capacity and fluid intelligence did not transfer to superior performance on the divergent thinking task. Second, confirming our prediction, generating uses was correlated with significantly lower activation in right ventral lateral prefrontal cortex (VLPFC) in the experimental than control group (figure 12.2). Third, gain in fluid intelligence mediated the link between training and activation in VLPFC, suggesting a mechanism linking WM training to neural function in a target task. The results of our study (Vartanian et al., 2013) demonstrate that WM training is correlated with neural efficiency in the brain during divergent thinking. Furthermore, they also reinforce the role of the lateral prefrontal cortex as the nexus where WM, fluid intelligence, and creativity intersect—at least in the case of divergent thinking.

Fostering Creativity

Based on a componential view of creativity, the focus of this chapter has been to demonstrate that the neural bases of creativity can be influenced by boosting WM and fluid intelligence—two of its components. However,

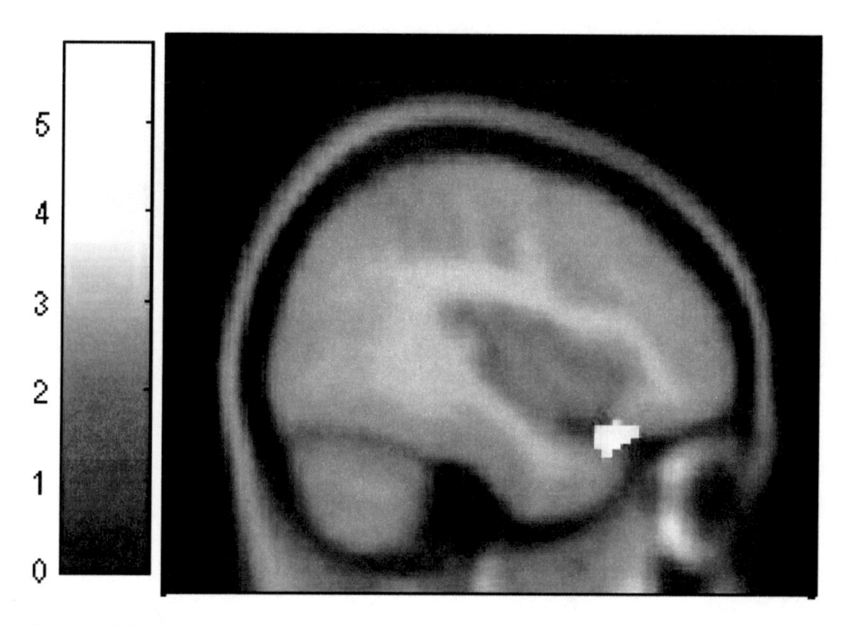

Figure 12.2
Impact of working memory training on neural function. Lower activation in right ventral lateral prefrontal cortex in participants who trained on the n-back task compared to the control group during the alternate uses task (see text).

tracking neural efficiency is one of several possible methods whereby neuroscientific evidence can be used to measure improvements in creativity and to use that knowledge to optimize training regimens to enhance it. For example, Fink and Neubauer (this vol.) review important studies in which computerized divergent thinking exercises were used for training creativity, addressing the research question of how brain activity may change as a function of creative thinking training. They found that the trained participants exhibited stronger task-related synchronization of frontal alpha activity than the control group. Such synchronization suggests that training may have served to make trained participants more focused on relevant aspects of the task. They also studied the impact of interventions (e.g., exposure to others' ideas) on creative production. Not only was this procedure effective in fostering creativity, but intervention was also reflected in activation in a network involving posterior brain regions known for their role in semantic information processing. Here, the simultaneous use of multiple imaging modalities to investigate whether increases in neural efficiency (by reductions in the BOLD response) are correlated with greater activity as measured by EEG would appear promising (see Fink et al., 2009).

Perhaps most promising is that neuroscience may help track transfer from creativity training to other noncreative tasks. For example, Forgeard, Winner, Norton, and Schlaug (2008) have shown that music training can benefit performance on activities distantly related to music, such as verbal ability and nonverbal reasoning. It is possible that such transfer is facilitated to the extent that the trained and output tasks recruit overlapping cortical regions, despite differences in surface features. In this sense, neuroscientific data can help elucidate the underlying mechanisms that may facilitate transfer not only from component processes to creativity, but from creative tasks to distant activities.

References

Amabile, T. A. (1985). Motivation and creativity: Effects of motivational orientation on creative writers. *Journal of Personality and Social Psychology, 48,* 393–399.

Amabile, T. A. (1998). How to kill creativity. *Harvard Business Review* (September): 77–87.

Aron, A. R., Robbins, T. W., & Poldrack, R. A. (2004). Inhibition and the right inferior frontal cortex. *Trends in Cognitive Sciences, 8,* 170–177.

Barch, D. M., Braver, T. S., Nystrom, L. E., Forman, S. D., Noll, D. C., & Cohen, J. D. (1997). Dissociating working memory from task difficulty in human prefrontal cortex. *Neuropsychologia, 35,* 1373–1380.

Buschkuehl, M., Jaeggi, S. M., & Jonides, J. (2012). Neuronal effects following working memory training. *Developmental Cognitive Neuroscience, 2S,* S167–S179.

Cattell, R. B. (1963). Theory of crystallized and fluid intelligence: A critical experiment. *Journal of Educational Psychology, 54,* 1–22.

Deary, I. J., Penke, L., & Johnson, W. (2010). The neuroscience of human intelligence differences. *Nature Reviews: Neuroscience, 11,* 201–211.

Duncan, J., Seitz, R. J., Kolodny, J., Bor, D., Herzog, H., Ahmed, A., et al. (2000). A neural basis for general intelligence. *Science, 289,* 457–460.

Fink, A., Grabner, R. H., Benedek, M., & Neubauer, A. C. (2006). Divergent thinking training is related to frontal electroencephalogram alpha synchronization. *European Journal of Neuroscience, 23,* 2241–2246.

Fink, A., Grabner, R. H., Benedek, M., Reishofer, G., Hauswirth, V., Fally, M., et al. (2009). The creative brain: Investigation of brain activity during creative problem solving by means of EEG and fMRI. *Human Brain Mapping, 30,* 734–748.

Fink, A., Grabner, R. H., Gebauer, D., Reishofer, G., Koschutnig, K., & Ebner, F. (2010). Enhancing creativity by means of cognitive stimulation: Evidence from an fMRI study. *NeuroImage, 52,* 1687–1695.

Fink, A., Koschutnig, K., Benedek, M., Reishofer, G., Ischebeck, A., Weiss, E. M., et al. (2010). Stimulating creativity via the exposure to other people's ideas. *Human Brain Mapping, 52,* 1687–1695.

Forgeard, M., Winner, E., Norton, A., & Schlaug, G. (2008). Practicing a musical instrument in childhood is associated with enhanced verbal ability and nonverbal reasoning. *PLoS ONE, 3,* e3566.

Gilhooly, K. J., Fioratou, E., Anthony, S. H., & Wynn, V. (2007). Divergent thinking: Strategies and executive involvement in generating novel uses for familiar objects. *British Journal of Psychology, 98,* 611–625.

Goel, V., & Vartanian, O. (2005). Dissociating the roles of right ventral lateral and dorsal lateral prefrontal cortex in generation and maintenance of hypotheses in set-shift problems. *Cerebral Cortex, 15,* 1170–1177.

Gray, J. R., Chabris, C. F., & Braver, T. S. (2003). Neural mechanisms of general fluid intelligence. *Nature Neuroscience, 6,* 316–322.

Hempel, A., Giesel, F. L., Caraballo, N. M. G., Amann, M., Meyer, H., Wüstenberg, T., et al. (2004). Plasticity of cortical activation related to working memory during training. *American Journal of Psychiatry, 161,* 745–747.

Jaeggi, S. M., Buschkuehl, M., Jonides, J., & Perrig, W. J. (2008). Improving fluid intelligence with training on working memory. *Proceedings of the National Academy of Sciences of the USA, 105,* 6829–6833.

Jaeggi, S. M., Buschkuehl, M., Jonides, J., & Shah, P. (2011). Short- and long-term benefits of cognitive training. *Proceedings of the National Academy of Sciences of the USA, 108,* 10081–10086.

Jung, R. E., & Haier, R. J. (2007). A Parieto-Frontal Integration Theory (P-FIT) of intelligence: Converging neuroimaging evidence. *Behavioral and Brain Sciences, 30,* 135–187.

Kim, K. H. (2005). Can only intelligent people be creative? *Journal of Secondary Gifted Education, 16,* 57–66.

Klingberg, T. (2010). Training and plasticity of working memory. *Trends in Cognitive Sciences, 14,* 317–324.

Li, S. C., Schmidek, F., Huxhold, O., Rocke, C., Smith, J., & Lindenberger, U. (2008). Working memory plasticity in old age: Practice, gain, transfer, and maintenance. *Psychology and Aging, 23,* 731–742.

McNab, F., & Klingberg, T. (2008). Prefrontal cortex and basal ganglia control access to working memory. *Nature Neuroscience, 11,* 103–107.

McNay, E. C., McCarty, R. C., & Gold, P. E. (2001). Fluctuations in brain glucose concentration during behavioral testing: Dissociations between brain areas and between brain and blood. *Neurobiology of Learning and Memory, 75*, 325–337.

Miller, L. A., & Tippett, L. J. (1996). Effects of focal brain lesions on visual problem-solving. *Neuropsychologia, 34*, 387–398.

Morrison, A. B., & Chein, J. M. (2011). Does working memory training work? The promise and challenges of enhancing cognition by training working memory. *Psychonomic Bulletin and Review, 18*, 46–60.

Neubauer, A. C., & Fink, A. (2009). Intelligence and neural efficiency. *Neuroscience and Biobehavioral Reviews, 33*, 1004–1023.

Neubauer, A. C., Fink, A. & Schrausser, D. G. (2002). Intelligence and neural efficiency: The influence of task content and sex on the brain–IQ relationship. *Intelligence, 30*, 515–536.

Neubauer, A. C., Grabner, R. H., Fink, A., & Neuper, C. (2005). Intelligence and neural efficiency: Further evidence of the influence of task content and sex on the brain-IQ relationship. *Cognitive Brain Research, 25*, 217–225.

Nusbaum, E. C., & Silvia, P. J. (2011). Are intelligence and creativity really so different? Fluid intelligence, executive processes, and strategy use in divergent thinking. *Intelligence, 39*, 36–45.

Owen, A. M., Hampshire, A., Grahn, J. A., Stenton, R., Dajani, S., Burns, A. S., et al. (2010). Putting brain training to the test. *Nature, 465*, 775–778.

Plucker, J. A., & Renzulli, J. S. (1999). Psychometric approaches to the study of human creativity. In R. J. Sternberg (Ed.), *Handbook of creativity* (pp. 35–61). New York: Cambridge University Press.

Poincaré, H. (1913). *The foundations of science*. Lancaster, PA: Science Press.

Ramsden, S., Richardson, F. M., Josse, G., Thomas, M. S., Ellis, C., Shakeshaft, C., et al. (2011). Verbal and non-verbal intelligence changes in the teenage brain. *Nature, 479*, 113–116.

Renzulli, J. S. (1986). The three-ring conception of giftedness: A developmental model for creative productivity. In R. J. Sternberg & J. Davidson (Eds.), *Conceptions of giftedness* (pp. 53–92). New York: Cambridge University Press.

Richmond, L., Morrison, A., Chein, J., & Olson, J. R. (2011). Older adults show improved everyday memory and attention via a working memory training regime. Poster presented at Annual Meeting of the International Neuropsychological Society, Boston, MA.

Schlaug, G., Forgeard, M., Zhu, L., Norton, A., Norton, A., & Winner, E. (2009). Training-induced neuroplasticity in young children. *Annals of the New York Academy of Sciences, 1169*, 205–208.

Sligh, A. C., Conners, F. A., & Roskos-Ewoldsen, B. (2005). Relation of creativity to fluid and crystallized intelligence. *Journal of Creative Behavior, 39,* 123–136.

Sternberg, R. J. (Ed.). (1999). *Handbook of creativity*. New York: Cambridge University Press.

Sternberg, R. J. (2008). Increasing fluid intelligence is possible after all. *Proceedings of the National Academy of Sciences of the USA, 105,* 6791–6792.

Vartanian, O., Jobidon, M-E., Bouak, F., Nakashima, A., Smith, I., Lam, Q., & Cheung, B. (2013). Working memory training is associated with lower prefrontal cortex activation in a divergent thinking task. *Neuroscience, 236,* 186–194.

Vartanian, O., Martindale, C., & Matthews, J. (2009). Divergent thinking is related to faster relatedness judgments. *Psychology of Aesthetics, Creativity, and the Arts, 3,* 99–103.

Wager, T., & Smith, E. (2003). Neuroimaging studies of working memory: A meta-analysis. *Cognitive, Affective, and Behavioral Neuroscience, 3,* 255–274.

Weisberg, R. W. (1999). Creativity and knowledge: A challenge to theories. In R. J. Sternberg (Ed.), *Handbook of creativity* (pp. 226–250). New York: Cambridge University Press.

13 The Means to Art's End: Styles, Creative Devices, and the Challenge of Art

Pablo P. L. Tinio and Helmut Leder

On the surface, the visual characteristics of artworks are similar to other objects in our surroundings. They may be described in terms of basic visual features such as shape, complexity, symmetry, color, and texture. However, people respond to and experience artworks in a different way. Although social factors certainly contribute to the special status of artworks (Gombrich, 1999), psychological factors play an important role as well. An artwork, unlike other objects in the environment, does not have a specific practical function—a brush paints, a spoon feeds, and a box holds; it may be argued that a painting is merely viewed. In this chapter, we argue that art does other things, although things that are more psychological than practical. Artworks, through specific means that artists employ, affect the viewer psychologically in a special way: they challenge and enlighten as well as pique and seize the interest of perceivers. This is made possible by creative means that artists have discovered and developed throughout the history of human art making.

Since Fechner's (1876) early studies, the psychology of aesthetics and the arts has suggested this special function of art. In this chapter, we provide an in-depth examination of the specific techniques and styles—the creative devices—that artists use to create artworks that not only correspond to the way we perceive and interpret the world (e.g., Gombrich, 1960; Zeki, 1999), but also challenge the psychological mechanisms that allow us to respond adaptively to the world. This perspective demands an examination into *both* sides of the experience of art, namely, creativity and aesthetics—what the artist does and what the viewer experiences. This interface is central to the special quality of art objects.

The way that an artwork challenges a perceiver depends on which aspect of cognitive processing is involved. Every creative device that an artist employs impresses on the viewer in a particular manner and affects specific psychological mechanisms—shape perception, object recognition, speed of information processing, implicit memory integration, interpretation and meaning-making, emotional reactions, and aesthetic judgments. To clarify these dynamics, recent conceptualizations of the aesthetic experience of art have placed particular emphasis on the different phases of art processing and corresponding psychological mechanisms (e.g., Chatterjee, 2003; Leder, Belke, Oeberst, & Augustin, 2004; Tinio & Leder, 2009). One approach, proposed by Leder et al. (2004), describes five stages of aesthetic experiences. During the early stages of aesthetic experiences, low-level visual features (e.g., symmetry) and modulating factors (e.g., familiarity) are processed. The perceiver's art-related knowledge becomes influential during the later stages, where the content and style of an artwork are determined and it is interpreted and evaluated. At any point during the process of art-viewing, different techniques and specific aspects of an artist's style could affect the outcome of the aesthetic experience—whether to enlighten us, pique our interest, or challenge how we normally see the world.

As part of their creative toolbox, artists have techniques that produce specific outcomes. The expression of these outcomes is, however, a process made complicated by a variety of factors. The very experience of art itself is modulated by the dynamic interplay between the characteristics of the art object (Kreitler & Kreitler, 1972) and its perceiver (Reber, Schwarz, & Winkielman, 2004), and the context in which this interplay is taking place. Another complicating factor is that similar techniques may be applied to different types of art, such as the juxtaposition of objects in both painting and photography. However, differences exist in their exact application and in the nature of their effects on the aesthetic experience. To account for such differences, which are often very subtle, we discuss artists' techniques and styles across various visual arts media, noting the differences as well as the similarities. It is important to note, however, that these discussions are also relevant to other nonvisual artistic media.

The approach of looking at art in terms of the challenges that it poses for the viewer is consistent with recent perspectives that endorse a broader conceptualization and examination of aesthetic experiences. For example, Silvia (2005, 2009) argued that aesthetics research is still closely tied to the tradition of Fechner (1876) and Berlyne (1971). As a consequence, aesthetics studies have focused mainly on liking and preference. According to Silvia, aesthetics should go beyond these positive responses and account for a broader range of aesthetic experiences, even negative ones such as disgust, anger, and confusion. Thus, the perspective in this chapter, which focuses on the psychological challenges posed by art, is consistent with this broader take on aesthetic experiences.

In our discussion of the challenge of art, we refer directly to studies of various psychological phenomena. At the same time, we also introduce new issues and propose novel concepts (e.g., the photographic peak-shift effect) that we believe are amenable to psychological testing and are particularly suited for, and relevant to, experimental and neuroscientific studies. In the next sections, we describe the following creative devices and their corresponding effects on specific aspects of aesthetic processing: line depiction and object recognition; figure-ground, juxtaposition, and visual continuation; the peak-shift effect and visual emphasis; sharpness of depiction and gaze patterns; framing and visual balance; physical orientation and size of an artwork and access to its content; and abstraction and speed of processing.

Line Depiction and Object Recognition

Line depiction is an artistic device that provides a definite psychological challenge to perceivers. This technique takes advantage of how humans process lines during the process of object recognition. Many theories of visual perception have claimed that object recognition proceeds through a series of representations that are constructed from the flow of visual information emanating from the environment. This process begins with the entry of photons into the eyes and results in rich perceptual representations occurring deep in the visual areas of the brain. David Marr (1982) attempted to systematically account for the sequential nature of visual perception by proposing a stage model that included very specific mechanisms and processes. As a starting point, he claimed that the visual system first analyzes the light distribution on the retina with the goal of identifying interesting visual areas. This process occurs via a filter that identifies areas of the image with transitions in contrast and where surfaces come together or overlap. Interestingly, and relevant to line depiction as a creative device, when this process is simulated in an artificial system, it produces scattered lines that are distributed over the image area and represents segments of the image that mark edges of objects as they appear in the real world. From this, Marr and others studying visual perception (e.g., Biederman, 1987) discussed how these kinds of lines could be the basis for object recognition. The researchers developed theories of object recognition based on the fundamental concept of basic elements, such as line segments, serving as input into the visual system and eventually resulting in a reconstruction of the visual environment.

If lines are the basic ingredients of vision, then it follows that perceiving objects depicted as line drawings should take place relatively easily; in fact, early studies (e.g., Biederman & Cooper, 1991) suggested that objects depicted using line drawings are, at the very least, not more difficult to perceive than the same objects depicted using visually richer media such as photographs. This is not surprising because of the importance of lines in major theories of vision (e.g., Marr, 1982). Many recognition skills, however, cannot be accomplished well from line representations. For example, figure 13.1 shows a transformation of a photograph to a line drawing using a line-extraction filter from a commercial image-processing software. Although the original scene can be identified based on the line depiction, it requires some effort—visual elements need to be combined to discern the boundaries of objects and major areas of the image in order to allow the proper identification of the scene. As a comparison, figure 13.2 is a photograph version of the same scene. It is apparent that scene information in the photograph is much richer and much easier to access than in the line depiction.

Figure 13.1
Landscape photograph converted to a line depiction (by P. L. Tinio).

Figure 13.2
Normal landscape photograph (by P. L. Tinio).

Face perception is one of the best examples of the effects of line drawings on perception. For example, it has been shown that line drawings showing faces are recognized as people but not as specific persons. Davies, Ellis, and Shepherd (1978) were the first to systematically study the difficulty of recognizing facial identities from line drawings of faces. Their research was motivated by the question of how images of faces could be produced in a way that facilitates the recognition of crime suspects and fugitives. The study compared people's ability to recognize simplified or detailed line drawings that were produced by accurately tracing the lines from corresponding original photographs. Compared to the photographs, for which accuracy of recognition was 90 percent, the accuracies for the line drawings were 47 percent for the detailed and 23 percent for the simple versions. This is evidence that line drawings are poor representations of facial identity. Leder (1996) conducted a series of studies with the goal of understanding why facial recognition is more difficult with line drawings. He found that the identification of factors such as age and attractiveness, which are typically performed with very little effort, were disrupted. Moreover, the configuration of facial features—information considered particularly important for face recognition—could not be accessed from line drawings.

From these studies, it is clear that the recognition of objects and faces suffers when they are depicted using lines. Line depictions also pose a challenge in perceiving artworks. One reason is that line depictions lack the shading information that is essential for object recognition (Davies et al., 1978). Shading provides cues—such as texture and depth—that are necessary for visual perception in general, and object recognition in particular. Artists manipulate shading both to facilitate and to hinder perception. Bruce, Hanna, Dench, Healey, and Burton (1992) have found that adding detail to the internal areas of line drawings eases their recognition. Artists know about these effects. Drawings using pen, charcoal, or pencil contain mainly lines. These are often handled using hatching to produce areas of texture that facilitate recognition. A close look at celebrated engravings, such as those by Albrecht Dürer, shows that they are rich with "fill" information.

These techniques are central to an artist's creative arsenal and they are basic skills that must be mastered in the first years of art school or professional apprenticeship. A source of evidence that artists facilitate, perhaps implicitly, object recognition in their artworks comes from the field of neurostatistics. Graham and Redies (2010) examined the image statistics of line drawings as a class of artworks. They used a large set of images and analyzed the Fourier power spectra of various classes of objects. Their analyses showed that the slope of the distribution—an indicator of the relationship between detail and gross areas of an image—for portrait drawings was more similar to the statistical values found for photographs of real-world scenes than, for example, photographic portraits. This suggests that artists have implicit knowledge that allows them to produce drawings with visual statistical characteristics that resemble visual stimuli for which the human visual system is presumably best adapted—natural scenes, which are the type of environments from which humans evolved.

In contrast to the types of depictions that feature techniques that attempt to ease recognition, some artworks seem to have been made to hinder recognition, thus challenging the perceiver. Andy Warhol's portrait paintings are good examples of this. Some of his most recognized portraits were of Marilyn Monroe, which were typical of his style of painting. These portraits combine the characteristics of line drawings (with the outline defining the shape, but lacking in internal details) and the characteristics of photographic negatives. It is known that facial identities in photographic negatives are difficult to recognize (Phillips, 1972). Warhol's novel use of color further disrupts aesthetic processing. Thus, the aesthetic challenges associated with such artworks are immense.

Figure-Ground, Juxtaposition, and Visual Continuation

The juxtaposition of objects—or the partial superimposing of an object over another causing the partial occlusion of the second object—is another means by which artists influence the aesthetic response of the perceiver. This creative device is an exploitation of one of the most fundamental features of visual perception: the ability to distinguish an object from its background, which is also referred to as the figure-ground phenomenon first described by the Gestalt psychologists (e.g., Koffka, 1935; Rubin, 1915/1958). Being able to separate an object of interest from its surroundings is a prerequisite for the identification or recognition of that object. The significance of this human ability is further evidenced by recent studies that have shown that figure-ground separation occurs even in the absence of directed attention to an object (Kimchi & Peterson, 2008), which also suggests the adaptive value of being able to separate an object from its surroundings.

The Dutch artist M. C. Escher takes advantage of the human predisposition to separate figure from ground. In his works, Escher creates what may be considered visual illusions in which visually separating one object from another, or an object from its background, becomes difficult. These works challenge perceivers by disrupting the visual perceptual responses that have been adapted to their usual environment. Escher's *Sky and Water II*, a grayscale woodcut, depicts birds on the top half and fish on the bottom half of the frame. The meeting of the two types of figures on the water line poses a figure-ground challenge: effort is required to distinguish the birds from the fish, and the perceiver is left with the task of alternating attention between the two groups of objects. The manipulation of figure-ground—whether done consciously or unconsciously—is central to Escher's style, and this may be the underlying reason for the engaging quality of his artworks.

Artists can use juxtaposition for visual effects in another way: through the continuation or discontinuation of the form of an occluded object. Arnheim (1954) provided the following example: "The segment of a disk will or will not appear as a part of a circular shape depending on whether the curvature, at the points of interruption, suggests continued extension or an inward turn toward closure" (p. 122). He presented a detailed discussion of the effects of juxtaposition on the perception of artworks. Arnheim believed that juxtaposition accentuates and makes the visual relationship among objects more dynamic. Separation of figure from ground through juxtaposition leads to an emphasis of one object over another—a visual hierarchy; although this is not always the case, as it could also be that there is an alternation of juxtaposition between the various elements of two objects within an image. An example of this would be a drawing of two snakes that are intricately intertwined; here, which snake occludes and which is occluded depends on at which part of the drawing one is looking.

In photography, the figure-ground relationship is also manipulated in specific ways to produce effects that challenge the perceiver. For example, the type of camera lens used to create a photograph affects the perceived relationship between the figure and background or between any number of objects occupying different visual planes within the image frame. Camera lenses with "long" focal lengths (telephoto lenses), in contrast to those with "short" focal lengths (wide-angle lenses), compress the image, and as a result, objects in different visual planes appear closer to each other (London & Upton, 1998), an effect similar to that produced by the side mirrors of cars where objects in the mirror appear closer than they actually are. In this sense, the photographer could produce certain effects by exploiting the relationship among objects in terms of depth and perspective. Perceivers are challenged because what they are seeing in a photograph does not correspond to how they typically see the world.

Juxtaposition of visual elements is prevalent in photography, and its effects on perception and the aesthetic experience are similar to its effects and experience in other art media. It is, however, important to note that overlapping in photography has effects that are specific to the medium. The uniqueness of the photograph as an image lies in its ability to capture a real scene, a "frozen" moment of an actual object in its environment. Although a photograph distorts how its referent is perceived (Latto & Harper, 2007), perceivers nonetheless *expect* it to have sufficient accuracy of depiction and represent an actual moment in time. Many photographs capture dynamic and multilayered scenes with objects moving in different directions. A photograph of such a scene will almost certainly have overlapping objects or moving *objects with overlapping trajectories*—with each object on a course of its own. Such an image is particularly engaging. Palmer, Gardner, and Wickens's (2008) recent study on the influence of the position and direction of an object on the spatial composition within a frame provides evidence of why this is the case. They found a preference for images depicting objects that are facing their direction of movement and that people have an *inward bias*—*inward* referring to objects that face into as opposed to out of the frame. For example, an image with an object facing left and positioned on the right side of the frame is preferred to the same object on the left side of the frame—the former object is inward facing while the latter is outward facing. This effect was found in a two-alternative forced-choice task and in an adjustment task in which participants placed objects on a horizontal line. Palmer et al. also employed a photograph production task where participants took photographs of objects. In the latter task, participants positioned objects facing left on the right side of the frame and objects facing right on the left side of the frame.

Based on such results, a person viewing an image depicting objects with different movement trajectories must juggle simultaneous expectations—an expectation corresponding to each moving object. Photographs have an element of truthfulness that other artistic media do not have. As a result of this, there is a natural tendency to want to know what happened before the depicted moment and to desire the completion of an action—an expectation of where a moving object is headed. Participants taking photographs in Palmer et al.'s (2008) study provided objects the space in which to move. People do not have similar expectations of depicted objects that do not have an implied direction—forward-facing objects, for instance. In fact, Palmer et al. found that participants preferred forward-facing objects to be positioned near the center of a frame.

For photographs with multiple objects, the different trajectories of the various objects overlap, and viewers, because of their expectations, must presumably expend effort to engage with the objects aesthetically. The result is deep engagement with the photograph; compare this to a still-life photograph of a flower. The dynamism and interestingness of a photograph with moving objects is thus rooted in the viewer's attempt to realize an expectation and a completion of a *visual aesthetic narrative*, which demands cognitive effort and facilitates deep engagement. Figure 13.3 is an example of a photograph that has a high demand for a completion of a visual aesthetic narrative. The viewer must constantly navigate between the figures that compose a network of overlapping trajectories.

What we have learned from the previous examples is that figure-ground manipulation and overlapping could influence perceivers in several ways and that artists could exploit this to entice and to challenge the perceiver. An object can be visually emphasized more than another. A qualitative dominance hierarchy between objects could also be produced, with some objects being perceived as dominant and some as submissive. Thus, the affective reactions of perceivers are highly susceptible to the effects of such manipulations. Arnheim (1954) provided the example of photographing a prisoner from either inside or outside his prison cell. As one could imagine, the two views would elicit different reactions from the perceiver. Artists could also employ juxtaposition to force perceivers of their artworks to

Figure 13.3
Photograph illustrating *objects with overlapping trajectories* and *completion of a visual narrative* (*Untitled* by P. L. Tinio).

alternate their focus between two objects that are entwined so as to engage them and enhance the dynamism of the artwork. Alternating between two aspects of an artwork could also lead to an increase in the ambiguity of the artwork, leaving perceivers uncertain as to how to approach an aesthetic encounter—their challenge is to resolve (or at least tolerate) the visual ambiguity. Juxtaposition may be used to create mystery where the main subject or primary action in a painting is occluded and the viewer must engage with the painting using the limited information available. Finally, juxtaposition in photographs of moving objects creates a demand for a completion of a visual aesthetic narrative where the different object trajectories each compete for attention and demand a resolution.

Peak-Shift Effect and Visual Emphasis

As much as artists have at their disposal techniques that slow, disrupt, or conceal a depicted object, they also have ways of isolating, highlighting, and exaggerating particular characteristics of an object. According to Ramachandran and Hirstein (1999), these latter techniques are exemplified by the *peak-shift effect* in art, which is one of the main underlying principles behind the human fascination for art. This effect also corresponds specifically to neural response patterns: amplification of form leads to a corresponding amplification of neural responses to such form. The caricature (see Gombrich, 1960, for a historical perspective) is one manner of artistic depiction that reflects the peak-shift effect. Caricature artists identify the most distinct aspects of a person's physical appearance and amplify them. For example, the television host Jay Leno's chin, when compared to chins of most people, is distinctively large and elongated. A caricature artist drawing his face would accentuate—by further enlargement and elongation—this already distinctive and idiosyncratic feature. Artists thus employ the peak-shift effect—deliberately or not—to emphasize particular aspects of their subject.

Ramachandran and Hirstein (1999) argued that caricature in one form or another is present in most art and is not limited to form space (i.e., the contour or shape of a visual object), but may also apply to other visual elements such as color. This would be the case with an artist who exaggerates the *blueness* of a blue sky or a blue ocean.

The peak-shift effect in aesthetics has been described primarily in the context of the traditional arts such as painting. Photography, however, is an art medium in which the peak-shift effect is highly prevalent. At the same time, it is a medium in which the effect is difficult to employ deliberately for the reason that most photographs, especially those depicting people, are created in mere hundreds of seconds (the short exposure time is necessary to "freeze" even the smallest of movements). Ramachandran and Hirstein (1999) described the peak-shift effect concerning body postures in terms of the following process: first, a given body posture is subtracted from an average of body postures (*posture space*) that a person has seen during a lifetime; second, the resulting difference is then amplified (i.e., exaggerated). From this an effect is produced where a less-than-average posture becomes even more distinct, but still resembles a known or even regularly observed posture.

We describe the peak-shift effect in photography as the process of capturing in an image a less-than-average body position, nonverbal gesture, or facial expression. This results in a powerful image—with universal appeal—that has more than average ability to affect the viewer. The most compelling documentary photographs are the best examples of this. As an example, consider Dorothea Lange's iconic photograph *Migrant Mother* (figure 13.4), made during the Great Depression, which represented the plight of poor migrant workers and their struggling families in the United States (Rosenblum, 2007). How could a single image have such a powerful effect? We argue that photography produces images that are optimal for evoking a strong response from the viewer through a *photographic peak-shift effect*.

The photographic peak-shift effect is rooted in the attributes of photography that distinguish it from other art media. First, cameras have the ability to freeze moments in time, especially moving objects, even when the movements are so brief that they are rendered nearly imperceptible to people. Second, photography is typically employed to document people in special situations (birthday parties and family gatherings), newsworthy occurrences (a political protest), and significant historical events (war). Thus, some body postures, nonverbal gestures, facial expressions, and movements, because they are imperceptible and rare, do not become part of the posture space, which comprises canonical representations of these human body states; however, although imperceptible and rare, they still resemble, even if only slightly, something in people's memory. In addition to the two previous reasons (imperceptibility and rarity), photography has a distinct quality that is absent in other visual arts media: the realism of content—photographs involve real people and their movement in space situated within real environments.

Lange's photograph is powerful because it captures a moment in which the mother's facial expression and nonverbal gesture is at its peak and this is accompanied by the positions of the children. According to the legendary

Figure 13.4
Dorothea Lange, *Migrant Mother*. Library of Congress, Prints & Photographs Division, FSA/OWI Collection, LC-USF34-9058-C.

and highly influential photographer Henri Cartier-Bresson, "You wait and wait and then finally you press the button—and you depart with the feeling (though you don't know why) that you've really got something. Later, to substantiate this, you can take a print of this picture, trace on it the geometric figures which come up under analysis, and you'll observe that, if the shutter was released at the decisive moment, you have instinctively fixed a geometric pattern without which the photograph would have been formless and lifeless" (1952, p. 8). The photographic peak-shift effect explains this decisive moment about which Cartier-Bresson speaks.

Painters and sculptors may rely on prototypical or canonical representations about facial expressions such as sadness and joy and about body postures such as those that reflect aggression and surprise. The photographer, in contrast, exploits peaks in nonverbal behavior that, because of the quickness in which they are expressed, render them imperceptible; photographic images of these peaks, which are more extreme than prototypical or canonical representations, elicit a strong response from the perceiver.

Sharpness of Depiction and Gaze Patterns

An additional visual element controlled by artists is the sharpness (or clarity) with which an object is depicted. Certain art media, such as aquarelle and pastel, produce artworks that have relatively low clarity. In a study examining the relationship between processing fluency and aesthetic preference, Leder (2003) presented participants with two versions of pastel portraits—one with low clarity and the other with moderately high clarity—and one aquarelle portrait that had extremely low clarity. Results showed that the stimuli that were higher in clarity, and thus higher in processing fluency, were preferred more. In this example, clarity of depiction was constrained by the actual choice of art materials.

Photography is another medium where the manipulation of clarity—through variation in sharpness—is highly evident. The nature of this manipulation is closely tied to two things. The first is the manner in which a camera captures an image. As complex as cameras have become during recent years, the essence of picture taking still comes down to the photographer's regulation of basic camera settings (lens focus, aperture opening, and shutter speed) that influence the sharpness of an image. The second is the characteristics of the perceiver.

For the human visual system, sharpness is related to where we need to look at any given moment. Only a small area—approximately two degrees of visual angle—of the field of vision is in focus. As you move away from this central region to the periphery, sharpness declines dramatically. The eyes must move constantly to scan different areas of the visual field (Rayner, 1998). Thus, when we view a photograph, our eyes must move constantly across the image, fixating on different areas. The photograph (and other static images) is visually interesting and provides the viewer with challenges that are rooted in the difference between the workings of the camera and human vision.

Because of physiological constraints, the mechanics of human vision is fixed—the process of scanning of a visual field is relatively similar across different contexts. This contrasts the wide range of sharpness of depiction across photographs: one photograph could have a very shallow depth of field with sharp focus on only one visual plane; another could have greater depth with more objects from different visual planes in sharp focus. These extremes in sharpness characteristics provide viewers with visual experiences that are similar and yet in some way different from the way they normally perceive the world. Photographs are able to depict real scenes fairly accurately and are thus close to how people see scenes in their daily lives; yet photographs are able to represent sharpness in a way that is very different from what the human visual system is able to achieve. For example, photographs that have extremely high levels of depth-of-field—that is, objects from the foreground to the background of the image are all relatively equally in sharp focus—are extremely unusual in terms of how people normally see the world, because in normal seeing, only a small portion of the visual field is in focus (Rayner, 1998). Thus, to a viewer, extremely sharp images, especially those depicting changing and moving objects, are visual treats. They allow inspection of a dynamic scene in a non-time-sensitive manner; in other words, through photographs, people are able to visually inspect a scene without the need to take into account changes that occur as time passes—the aesthetic experience of a static image is, in a way, time independent.

Framing, Cropping, and Visual Balance

Up to this point, we have discussed techniques that artists use that pose particular challenges to the viewer. These techniques involve manipulation—whether conscious or not—of the most essential aspects of an image typically involving the object (e.g., its sharpness or the juxtaposition of elements) being depicted. Artists, in addition, go to great lengths to control the physical boundaries or edges around the image. Modern imaging techniques, which enable art historians and conservators to analyze the process that an artist goes through to create a work of art, have shown that multiple revisions are often made on the boundaries of an image (see Kirsh & Levenson, 2000).

Controlling the boundaries of an image is particularly important in photography. Taking a photograph involves placing a rectangular frame around a live scene or object and permanently excluding everything outside that frame from the final image. This process of framing is unique to photography. For example, in painting, the artist begins with a blank canvas upon which various layers of marks and depictions are built, usually beginning with an underdrawing where the underlying structure of the final image is partially defined. The painter then continues to build the image. In contrast, the photograph begins with an existing image, which the photographer must refine through a process of selection. (Excluded in this discussion is, of course, photography that involves artificial selection and arrangement of scenes.)

The viewfinder of a camera, acting as a frame, provides a coherent visual organization to a world that actually lacks such organization, thus creating a sequence of representations (Burgin, 1977). The viewfinder thus selects and organizes the raw materials with which the photographer is working. This may be considered as the initial phase for establishing the composition of the image—a process not unlike putting down the underdrawing in a painting. As with other image manipulation procedures, cropping influences the viewer in specific ways. Tinio and Leder (2009), in their *taxonomy of image-manipulation procedures*, characterize the effects of cropping on the aesthetic experience of the viewer. Cropping is a *composition-level manipulation* that affects the structure of images. Cropping changes the perceived symmetry, complexity, and overall pictorial balance of images, characteristics that are processed during the early stages of information processing.

Recently, McManus et al. (2011) examined photographic cropping in terms of both how people crop photographs and how people aesthetically evaluate photographs that have been cropped in various ways. Their findings from six psychophysical studies showed that cropping indeed has a strong influence on people's aesthetic judgments. First, they found that people were consistent in terms of how they cropped different photographs. Second, they showed that some people are better croppers than others. Third, there were differences between the way art experts and nonexperts cropped photographs. These findings serve as evidence that both the act of cropping and the preference for a particular cropping style are not arbitrary. Photographic cropping is indeed a feature of both creative expression and aesthetic experiences.

A photographer's style may be defined by a preference for (or objection to) the use of cropping to finalize the composition of images. Cartier-Bresson (1952), for instance, vehemently objected to the cropping of his photographs, and believed that cropping leads to a disruption of the balance of a good photograph. Quite the opposite opinion was held by Alexey Brodovich (1961), who believed that cropping helps photographers interpret their photographs in new ways and that it should be employed if it will result in more powerful photographs. In terms of artistic styles related to cropping, these two artists represent the opposite ends of the spectrum, and their adherence to such styles further underscores the importance of cropping as a creative technique.

Physical Orientation and Size of a Work and Access to Content

A completely different kind of framing, one related to the physical orientation of an artwork, has also been used by artists to obtain specific effects. Except for a handful of studies (e.g., Latto & Russell-Duff, 2002), the orientation of an artwork has not been extensively discussed in the aesthetics literature. The manipulation of physical orientation is a very efficient way of influencing how an artwork is perceived. It can also be used as a creative way of posing a challenge to the viewer. The use of this method is prominent in the works of the German painter Georg Baselitz. A dominant feature of Baselitz's style is the upside-down presentation of his paintings, a method that he has employed since the 1970s. Viewers may notice that this mode of presentation has a dramatic effect on how the contents depicted in the paintings are accessed—that inversion disrupts the processing of content and transforms an image from representative to abstract. Although this claim might seem like an exaggeration, it is in accordance with findings showing that the inverted presentation of faces disrupts their processing—the so-called *face inversion effect*.

Yin (1969) was the first to assume that inversion would disrupt processing of objects in general, and that this effect would be strongest for the processing of faces. Since his seminal work, studies have revealed the psychological mechanisms underlying the effect: a kind of processing in which both holistic representation and reliance on configural information become unavailable when faces are inverted (e.g., Leder & Bruce, 2000). Baselitz's inverted artworks therefore target people's processing of orientation information and their need to know the orientation of up and down (e.g., ground and sky) as they relate to their surroundings. The question remains of how other regularities in our visual environment are disrupted by other atypical orientations of artworks. This question becomes even more complicated when physical orientation interacts with other visual elements such as color, which is the case with some abstract paintings. For example, what would be the effect of disrupting the often-found distribution of brown to blue (representing the transition from ground to sky) that has been identified as an important attribute of landscape images (Oliva & Torralba, 2001)?

Another important physical aspect of artworks is size. Artists make conscious decisions regarding the scale of their artworks relative to content and what they are trying to achieve. Renowned artists have created artworks on a very small scale. Examples include Vermeer's painting *The Lacemaker*, which measures approximately 9.6 × 8.3 inches, and some paintings by Max Ernst and Salvador Dalí. The diminutive sizes of such artworks bear no relation to their power to captivate perceivers or to their significance in terms of their place in art history. There are also many examples of artists creating noticeably large artworks. Examples include Pablo Picasso's *Guernica*, which measures 138 × 308 inches, and the color field paintings by abstract expressionist artists such as Jackson Pollock and Barnett Newman.

The influence of the size of an artwork on aesthetic experiences could also be modulated by what and how content is depicted. In a fascinating study of the influence of an artwork's size on the aesthetic experience, Pelli (1999) examined the works of Chuck Close, an American painter who creates large-scale works based mainly on dividing the canvas with a grid and filling in each cell of the grid with color. This style of depiction produces visual impressions that depend on a viewer's distance from the artwork: seen from near, each of the many individual cells is perceived, whereas seen from afar, small groups of cells meld together to form features of a larger image, a classic example of which is the portrait—one of the main aspects of his oeuvre. In several studies, Pelli (1999) directly referenced Close's artworks when examining the relationship between viewing distance—hence object size—and the perception of an object. Pelli's results demonstrate that aesthetic experiences are dependent on the sizes of objects; in this case, the size of an artwork is a function of the perceiver's distance from it—the faces in Close's artworks are only perceptible from afar.

Abstraction and Speed of Processing

Recent studies have provided evidence of a close relationship between characteristics of visual stimuli and the relative speed with which they can be processed. Visual perception happens so quickly that it is easy to assume that it is an instantaneous process. However, it involves brain activities and physiological processes that require time. Making estimates of the duration of early physiological processes is complicated because there is a strong interaction between visual processes and memory representations. As complex as visual perception is, it is a highly efficient process. Thus, even minimal visual information reaching the retina is interpreted, which results in fast percepts. Oliva and Torralba (2001) have shown that representations of what an image depicts rely on global image features. For example, an image with a light upper portion and a dark lower portion is perceived as depicting a landscape, and that an image with multiple green elements is perceived as depicting a forest. Such inferences have been shown to have good validity. Moreover, Torralba (2009) found that the content of an image could be extracted even if the resolution is as low as 32×32 pixels. Results of studies employing tachistoscopic image presentation—allowing millisecond stimulus presentation—have shown adequate recognition performance, with results showing that when images are flashed briefly into the left and right visual fields, perceivers can detect animals within 120 milliseconds (Kirchner & Thorpe, 2006) and faces within 100 milliseconds (Crouzet, Kirchner, & Thorpe, 2010).

As important as temporal factors are to visual perception, there is a lack of systematic studies of the time frame of art perception. Leder et al. (2004) attempted to account for the time frame of aesthetic experiences in their model of aesthetic experiences. Using this model as a foundation, Augustin, Leder, Hutzler, and Carbon (2008) recently examined the time it takes to process an artwork's content or style. Their results showed that artistic style, such as that of van Gogh, was processed quickly (after 50 ms); but not as quickly as what was depicted in the image. Bachmann and Vipper (1983) also examined the time course of aesthetic experiences by presenting artworks that featured six highly distinctive artistic styles. Results showed that the longer participants viewed an artwork, the more they tended to perceive it as more involved, simple, regular, and precise. These studies shed light on the time course of aesthetic experiences of visual art. However, they do not help determine the process by which an artwork—and its specific techniques and styles—directly affect the time frame of a viewer's aesthetic experience.

In art perception, certain devices that artists use have an impact on speed of processing. A good example is cubism, an artistic style that has been shown to influence the time frame of aesthetic experiences. Cubist paintings have the effect of slowing down information processing. In cubism, depicted objects are deconstructed into abstracted forms. Common to cubist paintings is the breakdown of an object into basic geometric elements, which interestingly resemble the forms in Biederman's (1987) theory of object recognition in which objects are represented as *geons*—simple forms such as cylinders and cones. In cubist paintings, the basic elements are then recombined to simultaneously depict multiple views, thus providing a kind of integrated and visually dynamic metarepresentation of an object. This mode of depiction involving the abstraction of image content and depiction of multiple views has been shown to slow the process of object recognition (Hekkert & van Wieringen, 1996). Recently, Kuchinke, Trapp, Jacobs, and Leder (2009) further explored the aesthetic response to cubist paintings using physiological methods and found that high abstraction in cubist paintings was associated with slower object recognition, more difficult processing, and more negative aesthetic judgments. They also found evidence that ease of processing was related to positive affect. As a creative device, abstraction has a tremendous impact on the time course of aesthetic experiences, and artists can modulate it to elicit positive or negative aesthetic reactions.

Conclusion

Art challenges. It does so in various ways depending on the creative device the artist employs—whether implicitly or explicitly. Each device targets a specific aspect of aesthetic processing. Line depictions hinder object recognition, which runs counter to some theories of visual perception that stress the importance of lines in object recognition (e.g., Marr, 1982). Juxtaposition influences the perception of figure-ground separation. It also suggests visual continuation and creates a dynamic effect when it involves objects with overlapping trajectories—viewers expect objects in motion to have a direction, and perhaps even a destination, that extends beyond the confines of the image. This becomes particularly challenging for viewers because such images demand a completion of a visual narrative. The peak-shift effect provides visual emphasis (Ramachandran & Hirstein, 1999). When applied to photography, the effect breaks free from its dependence on canonical representations stored in memory and involves gestures, postures, and facial expressions that are either so rare or nearly imperceptible that they can be experienced directly only when captured and frozen as a static image. Sharpness of depiction plays with gaze patterns because of the disparity in the way we look at an image of a scene and the way we look at the same scene in actuality. Cropping techniques affect the structure of artworks (Tinio & Leder, 2009). The physical orientation and scale of an artwork affects the way its content is accessed; an upside-down artwork is a different artwork right-side-up, and an artwork seen from near is visually different when seen from afar. Finally, visual abstraction slows information processing as the viewer attempts to determine the identity of the depicted object. The devices that we have highlighted above make up only a subset of what is available, of what has been discovered or developed during the long course of art making by humans.

Considering artworks in this way does have its limitations. Some of the devices, such as visual abstraction and line depiction, and their corresponding psychological effects, such as speed of processing and object recognition, pertain mainly to the early stages of aesthetic processing and only partially address the later stages where higher-order cognitions like meaning-making occur—the essence of the experience of art (Millis, 2001; Russell, 2003; Russell & Milne, 1997). This is not to say that the early stages are unimportant. Quite the opposite: what happens in the earlier stages set the stage for what happens in the later stages and what final outcomes subsequently emerge—an aesthetic evaluation, for instance. Moreover, we have not addressed the interactions among the many devices, which is likely to occur in the real world of art. For example, juxtaposition doubtlessly has an impact on the visual balance of an image, and decreasing the clarity of an image may also have the effect of increasing abstraction. There is also evidence that the effects of multiple image manipulations are additive (Tinio, Leder, & Strasser, 2010). The methodological challenges and conceptual ambiguities resulting from these interactions must be taken into consideration in future research.

Moreover, although our focus here was on the challenge of art, we do not discount the idea that aesthetic experiences involve pleasure. We do believe that some of this pleasure is derived from the resolution of the challenges imposed on the viewer, that an aesthetic experience is sometimes about problem solving—the type of problem determined by the device encountered by the viewer. This is consistent with recent models of the aesthetic experience (e.g., Leder et al., 2004). In the end, we believe that the power of art, its ubiquity in our lives, lies as much in its ability to challenge the way we look at reality as in its ability to represent and even mimic reality. Our approach is consistent with the recent push toward looking at aesthetics more broadly, beyond liking and preference (Silvia, 2005), and more toward the full richness of human experiences.

This approach coincides with developments in the ability to measure psychological responses (e.g., fMRI)—especially aesthetic responses. Thus, there is a correspondence between developments in the conceptualization of aesthetic responses and the use of neuroscientific methodologies. Data from imaging studies could be used to show the unique impact of artworks on viewers. However, neuroaesthetics is still an emerging subfield of aesthetics, and relatively few studies have been conducted on neural specificity associated with aesthetic responses to artworks. More importantly, there is a lack of neuroscientific studies that have examined nonpositive responses to art (Brown, Gao, Tisdelle, Eickhoff, & Liotti, 2011). The question of whether cortical structures that respond to artworks are the same as those that respond to everyday objects remains unresolved and open for examination. Consistent with our analysis of how artworks challenge the viewer is the idea that assessing a wider range of responses increases the possibility of discovering neural areas that are specific to artworks. Such an approach would go beyond simple perceptual responses common to most objects. It would take into account higher-order cognitive processes including meaning-making, art-related interpretations (favorable or not), judgments based on perceivers' art-related knowledge and experiences, and aesthetic emotions, both positive and negative. This would augment and build on knowledge gained from a century of work in experimental aesthetics, work that has mainly focused on the impact of simple, bottom-up factors on generally positive aesthetic responses (Silvia, 2005).

Looking at art in terms of challenges also provides a direct line between creativity and aesthetics. We believe that this has not been emphasized enough in the literature. Why is a focus on the challenges that art poses to the viewer better at connecting these two sides? Artists' motivations in creating artworks are not always based on whether viewers will find their creations pleasant; they consider other ways their creations could be received. The viewer's perspective is analogous to this. A visit to the museum is not just about pleasure, but also about getting interested and engaged, and it may even be about the process of confrontation—between a viewer's expectations and an artwork that is inconsistent with them. A visit to the museum could also involve exploration and the process of making meaning of an art object, both of which relate to the artist's perspective while creating the artwork. During aesthetic encounters, art viewers want to learn about the works, know about the motivations of the artists, and connect in some way to the ideas behind the works (Tinio, Smith, & Potts, 2010). The devices that artists use influence the outcomes of these encounters. Thus, looking at how the use of specific artistic devices challenge the viewer would bring creativity and aesthetics—creation and reception—closer to each other.

References

Arnheim, R. (1954). *Art and visual perception.* Berkeley, CA: University of California Press.

Augustin, M. D., Leder, H., Hutzler, F., & Carbon, C. C. (2008). Style follows content: On the microgenesis of art appreciation. *Perception, 128,* 127–138.

Bachmann, T., & Vipper, K. (1983). Perceptual rating of paintings from different artistic styles as a function of semantic differential scales and exposure time. *Archiv fur Psychologie, 135,* 149–161.

Berlyne, D. E. (1971). *Aesthetics and psychobiology.* New York: Appleton-Century-Crofts.

Biederman, I. (1987). Recognition-by-components: A theory of human image understanding. *Psychological Review, 94,* 115–147.

Biederman, I., & Cooper, E. E. (1991). Priming contour-deleted images: Evidence for intermediate representations in visual object recognition. *Cognitive Psychology, 23,* 393–419.

Brodovich, A. (1961). Brodovitch on photography. In C. H. Traub, S. Heller, & A. B. Bell (Eds.), *The education of a photographer* (pp. 133–139). New York: Allworth Press.

Brown, S., Gao, X., Tisdelle, L., Eickhoff, S. B., & Liotti, M. (2011). Naturalizing aesthetics: Brain areas for aesthetic appraisal across sensory modalities. *NeuroImage, 58,* 250–258.

Bruce, V., Hanna, E., Dench, N., Healey, P., & Burton, M. (1992). The importance of "mass" in line drawings of faces. *Applied Cognitive Psychology, 6,* 619–628.

Burgin, V. (1977). Looking at photographs. In L. Wells (Ed.), *The photography reader* (pp. 130–137). New York: Routledge.

Cartier-Bresson, H. (1952). *The decisive moment.* New York: Simon & Schuster.

Chatterjee, A. (2003). Prospects for a cognitive neuroscience of visual aesthetics. *Bulletin of Psychology and the Arts, 4,* 55–60.

Crouzet, S. M., Kirchner, H., & Thorpe, S. J. (2010). Fast saccades toward faces: Face detection in just 100 ms. *Journal of Vision, 10,* 1–17.

Davies, G., Ellis, H., & Shepherd, J. (1978). Face recognition accuracy as a function of mode of representation. *Journal of Applied Psychology, 63,* 180–187.

Fechner, G. T. (1876). *Vorschule der Ästhetik.* Leipzig: Breitkopf & Härtel.

Gombrich, E. H. (1960). *Art and illusion: A study in the psychology of pictorial representation.* Princeton, NJ: Princeton University Press.

Gombrich, E. H. (1999). *The uses of images: Studies in the social function of art and visual communication.* London: Phaidon Press.

Graham, D. J., & Redies, C. (2010). Statistical regularities in art: Relations with visual coding and perception. *Vision Research, 50,* 1503–1509.

Hekkert, P., & van Wieringen, P. C. W. (1996). Beauty in the eye of expert and non-expert beholders: A study in the appraisal of art. *American Journal of Psychology, 109,* 389–407.

Kimchi, R., & Peterson, M. A. (2008). Figure-ground segmentation can occur without attention. *Psychological Science, 19,* 660–668.

Kirchner, H., & Thorpe, S. J. (2006). Ultra-rapid object detection with saccadic eye movements: Visual processing speed revisited. *Vision Research, 46,* 1762–1776.

Kirsh, A., & Levenson, R. S. (2000). *Seeing through paintings: Physical examination in art historical studies.* New Haven, CT: Yale University Press.

Koffka, K. (1935). *Principles of Gestalt psychology.* New York: Harcourt Brace Jovanovich.

Kreitler, H., & Kreitler, S. (1972). *Psychology of the arts.* Durham, NC: Duke University Press.

Kuchinke, L., Trapp, S., Jacobs, A. M., & Leder, H. (2009). Pupillary responses in art appreciation: Effects of aesthetic emotions. *Psychology of Aesthetics, Creativity, and the Arts, 3,* 156–163.

Latto, R., & Harper, B. (2007). The non-realistic nature of photography: Further reasons why Turner was wrong. *Leonardo, 40,* 243–247.

Latto, R., & Russell-Duff, K. (2002). An oblique effect in the selection of line orientation by twentieth century painters. *Empirical Studies of the Arts, 20,* 49–60.

Leder, H. (1996). Line drawings of faces reduce configural processing. *Perception, 25,* 355–366.

Leder, H. (2003). Familiar and fluent! Style-related processing hypotheses in aesthetic appreciation. *Empirical Studies of the Arts, 21,* 165–175.

Leder, H., Belke, B., Oeberst, A., & Augustin, D. (2004). A model of aesthetic appreciation and aesthetic judgments. *British Journal of Psychology, 95,* 489–508.

Leder, H., & Bruce, V. (2000). When inverted faces are recognized: The role of configural information in face recognition. *Quarterly Journal of Experimental Psychology, 53A,* 513–536.

London, B., & Upton, J. (1998). *Photography.* New York: Addison-Wesley.

Marr, D. (1982). *Vision.* San Francisco: W. H. Freeman.

McManus, I. C., Zhou, F. A., l'Anson, S., Waterfield, L., Stöver, K., & Cook, R. (2011). The psychometrics of photographic cropping: The influence of colour, meaning, and expertise. *Perception, 40,* 332–357.

Millis, K. (2001). Making meaning brings pleasure: The influence of titles on aesthetic experiences. *Emotion, 1*, 320–329.

Oliva, A., & Torralba, A. (2001). Modeling the shape of the scene: A holistic representation of the spatial envelope. *International Journal of Computer Vision, 42*, 145–175.

Palmer, S. E., Gardner, J. S., & Wickens, T. D. (2008). Aesthetic issues in spatial composition: Effects of position and direction on framing single objects. *Spatial Vision, 21*, 412–449.

Pelli, D. G. (1999). Close encounters—an artist shows that size affects shape. *Science, 285*, 844–846.

Phillips, R. J. (1972). Why are faces hard to recognize in photographic negative? *Perception & Psychophysics, 12*, 425–426.

Ramachandran, V. S., & Hirstein, W. (1999). The science of art: A neurological theory of aesthetic experience. *Journal of Consciousness Studies, 6*, 15–51.

Rayner, K. (1998). Eye movements in reading and information processing: 20 years of research. *Psychological Bulletin, 124*, 372–422.

Reber, R., Schwarz, N., & Winkielman, P. (2004). Processing fluency and aesthetic pleasure: Is beauty in the perceiver's processing experience? *Personality and Social Psychology Review, 8*, 364–382.

Rosenblum, N. (2007). *A world history of photography*. New York: Abbeville Press.

Rubin, E. (1915/1958). Figure and ground. In D. C. Beardslee & M. Wertheimer (Eds.), *Readings in perception* (pp. 194–203). Princeton, NJ: Van Nostrand.

Russell, P. A. (2003). Effort after meaning and the hedonic value of paintings. *British Journal of Psychology, 94*, 99–110.

Russell, P. A., & Milne, S. (1997). Meaningfulness and hedonic value of paintings: Effects of titles. *Empirical Studies of the Arts, 15*, 61–73.

Silvia, P. J. (2005). Emotional responses to art: From collation and arousal to cognition and emotion. *Review of General Psychology, 9*, 342–357.

Silvia, P. J. (2009). Looking past pleasure: Anger, confusion, disgust, pride, surprise, and other unusual aesthetic emotions. *Psychology of Aesthetics, Creativity, and the Arts, 3*, 48–51.

Tinio, P. P. L., & Leder, H. (2009). Natural scenes are indeed preferred, but image quality might have the last word. *Psychology of Aesthetics, Creativity, and the Arts, 3*, 52–56.

Tinio, P. P. L., Leder, H., & Strasser, M. (2010). Image quality and the aesthetic judgment of photographs: Contrast, sharpness, and grain teased apart and put together. *Psychology of Aesthetics, Creativity, and the Arts, 5*, 165–176.

Tinio, P. P. L., Smith, J. K., & Potts, K. (2010). The object and the mirror: The nature and dynamics of museum tours. *International Journal of Creativity and Problem Solving, 20*, 37–52.

Torralba, A. (2009). How many pixels make an image? *Visual Neuroscience, 26*, 123–131.

Yin, R. K. (1969). Looking at upside-down faces. *Journal of Experimental Psychology, 81*, 141–145.

Zeki, S. (1999). *Inner vision: An exploration of art and the brain*. New York: Oxford University Press.

Contributors

Baptiste Barbot
Department of Psychology, Pace University

Mathias Benedek
Institute of Psychology, University of Graz

David Q. Beversdorf
University of Missouri

Aaron P. Blaisdell
Department of Psychology, University of California, Los Angeles

Margaret A. Boden
University of Sussex

Dorret I. Boomsma
Department of Biological Psychology, VU University Amsterdam

Adam S. Bristol
Aquilo Capital Management

Shelley Carson
Harvard University

Marleen H. M. de Moor
Department of Biological Psychology, VU University Amsterdam

Andreas Fink
Institute of Psychology, University of Graz

Liane Gabora
University of British Columbia

Dennis Garlick
Department of Psychology, University of California, Los Angeles

Elena L. Grigorenko
Yale University, Moscow State University, and Columbia University

Richard J. Haier
Mind Research Network

Rex E. Jung
Mind Research Network

James C. Kaufman
California State University—San Bernardino

Helmut Leder
Department of Psychology, University of Vienna

Kenneth J. Leising
Department of Psychology, Texas Christian University

Bruce L. Miller
Memory and Aging Center, University of California, San Francisco

Apara Ranjan
University of British Columbia

Mark Patrick Roeling
Department of Biological Psychology, VU University Amsterdam

W. David Stahlman
Department of Psychology, University of Mary Washington, Fredericksburg, Virginia

Mei Tan
Yale University

Pablo P. L. Tinio
Queens College of the City University of New York, Division of Education

Oshin Vartanian
University of Toronto—Scarborough

Indre V. Viskontas
Memory and Aging Center, University of California, San Francisco

Dahlia W. Zaidel
University of California, Los Angeles

Index